The Planning Guide to Piping Design

PROCESS PIPING DESIGN HANDBOOK

Volume One: *The Fundamentals of Piping Design*

Volume Two: *Advanced Piping Design*

Volume Three: *The Planning Guide to Piping Design*

The Planning Guide to Piping Design

Richard J. Beale, C.Tech
Association of Science and Engineering Technology Professionals of Alberta (ASET)

Paul Bowers
American Society of Mechanical Engineers (ASME)
Society of Piping Engineers and Designers (SPED), co-president
Calgary

Peter Smith

Gulf Publishing Company
Houston, Texas

The Planning Guide to Piping Design

Gulf Publishing Company
2 Greenway Plaza, Suite 1020
Houston, TX 77046

Library of Congress Cataloging-in-Publication Data

Beale, Richard, 1955-
The planning guide to piping design / Richard Beale, Paul Bowers, and Peter Smith.
p. cm.
Includes indexes.
ISBN-10: 1-933762-37-3
ISBN-13: 978-0-12-810226-8 1. Pipelines—Design and construction. I. Bowers, Paul (Paul Andrew), 1962- II. Smith, Peter, 1951 June 7- III. Title.
TA660.P55B43 2010
621.8'672—dc22
2010019043

ISBN: 978-1-933762-37-1

Printed in the United States of America

Editing, design and composition by TIPS Technical Publishing, Inc

10 9 8 7 6 5 4 3 2 1

To my mother and father who dedicated their lives to the raising of a family who loves, cares and has patience for one another, and to all of my family, who each in their own way has supported, encouraged and inspired me, I affectionately dedicate this book to all of you. Mum and Dad thank you for your hard work and wisdom. You have left a legacy to be proud of.

—Richard J. Beale

To my mother (for encouragement), grandfather (for teaching me to draw from an early age) and son (for inspiration). Also, thanks to all those engineers and designers who took the patience to inform and educate me over these past many years.

—Paul Bowers

To The Smith Family and to my son Stewart Smith. He is a constant source of inspiration and motivation to me. He is sadly missed by his family and friends, but he is never far away.
Stewart Smith, Musician. March 11, 1972–October 21, 2005

—Peter Smith

Contents

Foreword

Over the last two decades, process plant design has undergone two distinct changes. First, process plant design has adopted the 3D CAD model as the general data of record for plant equipment and piping layout. Second, plant Engineering, Procurement, and Construction (EPC) has adopted the Earned Value Management (EVM) approach to measuring project progress. We have needed to understand what impact this has had on plant equipment and piping layout. Richard Beale and Paul Bowers have finally addressed this need with this book. In it they explain how to estimate, plan, and execute a model based piping project as a "work package" managed enterprise.

The authors understand that 3D models have shifted the piper's role from drafter to data manager. In this high value added world, models are built, assured, and reconciled as the end product of design. Drawings, by contrast, are rendered from models and other databases. The answer to low labor cost competition is to greatly reduce or eliminate all labor costs. This is precisely what CAD and other automation tools provides us. These authors teach us how to plan jobs where multiple downstream activities (Procurement, Fabrication, Erection, etc.) must repurpose the work recorded in a small set of electronic models and databases.

It is in the second area, EVM, where Beale and Bowers add a new contribution to piping design. EVM lets projects measure progress by accumulating the value of the "work packages" they complete. The authors show us how to structure piping designs as packages for the designer, the fabricator, and the constructor. By planning for these discrete scopes of work from the beginning, focus remains on minimizing capital costs while maximizing functional performance.

As management guru Stephen Covey says, "Begin with the end in mind." This is what Beale and Bowers have accomplished for piping. Richard J. Beale has an enviable career spanning EPC piping design to owner/operator project planner. Paul Bowers compliments this with a full scope, in-depth design resume down at the lowest level of detail. Together, they provide an illustrated game plan for modern process plant project execution that should guide the young and back check the old.

—William G Beazley, PhD
Execution Director
Society of Piping Engineers and Designers

Preface

During the Christmas to New Year's period of the years 2007 and 2008, Paul Bowers and I had the honor to peer review the book *Advanced Piping Design* by Rutger Botermans and Peter Smith. While we were corresponding with Peter Smith a discussion developed about the need for a book that would provide guidance in the set-up and management of the piping design—a book for the project engineer or project manager who wishes to know more about the management of piping design and the piping designer who is new to the role of a piping lead.

There are many fine books available on piping design and plant layout, but we shared a common observation that there is very little information available on how to manage the piping design. It was from these initial discussions and our observation that the making of a book was born, the final result of which is this book that you hold in your hands today.

Essentially there are two components to the successful completion of the piping design: the layout of the piping systems and the management of the activities that support the layout of the piping systems.

Managing the piping design activities means being in control and organizing the planning, scheduling, manpower, progress monitoring, and communications of the piping design.

A piping designer begins his/her career after graduating from college, whereas a piping lead can only begin his/her career after many years of further study and experience gained by working as a piping designer. The knowledge and the abilities required to manage the activities that support the layout of the piping systems from conception through to construction can only be gained through years of exposure to piping design in organized companies with comprehensive procedures and through the guidance of experienced mentors.

A book cannot substitute for an experienced mentor, but this book will assist the piping lead by explaining the management side of piping design with reference to documented procedures. Because procedures provide direction to all team members and are integrally linked to the management of a successful piping design effort, they are the cornerstones of efficiency and teamwork. This book would have been incomplete without the inclusion of procedures. Until now documented procedures have only been available in company procedural manuals. Many of the procedures presented in this book have been used in industry for many years and are universally accepted because of the proven benefits, albeit with some differences to suit company specific requirements. For instance, a stick file procedure is an old concept used by many companies and has its origins going back many decades.

The methods of management and the procedures outlined in this book are recommended practices. It is not expected that you will agree with everything that is written here; however, it is not an option for a piping lead to ignore the management of an activity. The only choice is to select one method of management over another. It is compulsory for a piping lead to manage all of the activities in one way or another.

While the technologies used for piping design and drafting have changed from manual techniques to CAD over the last couple of decades, one thing has not changed: that of the need to consider all of the detail involved and manage accordingly.

During my career I have had the good fortune to work for some very senior piping leads and well-organized companies. I am grateful to all for teaching me the importance of attention to detail in the management of the piping design. I must give special thanks and credit to the leads and management of Delta Projects, Calgary, Alberta, Canada (now Jacobs) who I had the privilege to work for in my early years: because they, above all, taught me the principles and values that I carry with me to this day. Many of these people are now retired and this book is my way of continuing their legacy and passing on the convictions that they instilled within me to the next generation.

Credit and thanks are also due to the following individuals for their assistance and contributions to the writing of this book: Ross Krill, P.Eng., Team Lead, and Bob Baker, Document Control Group Lead, for providing the approvals for needed materials. Without their ongoing support the task of writing this book would have been much more difficult. Thanks also go to Xenia Beale, Intermediate Piping Designer; Mark Beauliua, Piping Lead; Madelaine Carrette, Contract Administrative Assistant; Gudrun Dahle, Senior Material Controller; Scott Maguire, Construction Coordinator; Rory McDougall, Manager of Pepco Pipe Services; Gord Mernickle, Construction Coordinator; Curtis Smith, Piping Lead; Alvin Winestock, P.Eng., PMP; and Sean Williams, CAD Administrator, for their willingness to give their time and expertise. Their assistance and advice has made a significant contribution.

It is my hope that this book will communicate and promote the importance of the management of piping design and provide guidance on how to do so.

—Richard J. Beale

One unintended consequence of the adoption of CAD technology has often been the fragmentation of piping designers' training, knowledge and experience into which software is being used. This is a natural outcome of the new tools being used and explored. Thirty years ago it did not matter if one used a Staedtler or Koh-I-Noor. Today's CAD tools are remarkable in their ability to facilitate 3D visualization, import/export data for engineering, procurement and construction purposes and to replicate designs. But it is important to recognize that machines only do what you tell them to do, and understanding in more detail the purpose of what you are "drawing" helps to create better designs and better designers. It is my hope that this book will be of interest to new piping designers/draftsmen/engineers and anyone else wanting to learn about the process of designing a process piping facility.

I wish to give thanks to the following people who helped me through my formative years: Glenn Fair, David Gencher, Robert Giubilei, Rene Inic, Bob Jacobs, Kim Levan, Steve Murton, Louis Parant, and Mike Schum.

—Paul Bowers

This book focuses on the management of the activities that support the piping design. It is our hope that it will prove to be a valuable resource to everyone interested in the role of a piping lead.

The experienced reader will notice that the focus of this book is on Engineering, Procurement, and Construction (EPC) and detailed design. This is because a piping lead will spend eighty to eighty-five percent of his/her time in detailed design, fabrication, and construction, where the true abilities of a piping lead are tested. The roles and responsibilities of the piping lead during Pre-FEED and FEED have not been discussed; however, it is emphasized many times in the book that one of the primary responsibilities of the piping lead is to know the project scope. Once a piping lead knows the project scope and his/her project deliverables it is a simple matter for him/her to extrapolate the information and guidance provided in this book that are relevant to his/her own situation.

The experienced reader will also notice the clear connection to the manual design and drafting world. This is because there are many small- to medium-size projects executed by small- to medium-size companies effectively utilizing manual procedures. Much of the work executed by these companies is somewhere between the realms of the manual and the automated due to the costs of software investment and/or clients who are not convinced of the benefits of using new technology on their projects. Often too, the larger projects and larger companies will fall somewhere in between the realms of the manual and automated depending on the software manufacturer of choice and the level of understanding of the available functionality of the software.

A book that reaches out to the global piping community who work in a variety of industries and with a variety of software must draw lines somewhere. The lines in this book are drawn at the median approach to piping execution between manual and automation where the vast majority of projects lie today. A key point for all readers is that the piping lead's roles and responsibilities do not change with technology; it is merely the methods of piping execution that may change. An understanding of where we have been, where we are currently, and where we are going is crucial for a piping lead to evaluate the effects of technological change and adjust accordingly.

Undoubtedly the path forward is one toward more automation and will become the subject of future discussions within all industries that involve piping design and all technical institutions that teach piping design. For instance, there is software already available capable of automatic pipe routing that reduces the time required for piping layout study and material take-off, but as of now this is not mainstream.

Your feedback will be appreciated so that we may make improvements. Feedback may be left with the Gulf Publishing Company at http://www.gulfpub.com or on the authors' website http://pipingdesign.com/planning_guide_piping_design/.

List of Figures

List of Tables

CHAPTER 1

Before You Begin

1.1 Introduction

In order to execute a project efficiently, it is essential that you plan your project. In order to do this you must first recognize all the questions that must be asked and answered, assemble all the needed tools, and make decisions. This chapter provokes thought: it focuses on the questions to ask and the tools required in order to begin a project. Do you have everything you need to proceed?

To begin, you must first assemble and then make yourself familiar with the engineering company and/or client standards, specifications, and procedures to be used on the project. Larger clients will have certain requirements in place and mandate that those requirements be used on the project, whereas smaller clients will likely default entirely to the engineering company. Generally speaking, all projects will use a combination of engineering company and client standards, specifications, and procedures. You must ensure that you know which you are using, and where they come from. As a piping lead it is doubly important to familiarize yourself with these requirements, not just so you can guide your team, but because you will likely have to explain them to other departments and insist that they respect and adhere to them.

As you progress in your career you will find that knowledge of these tools is required for any project, and the valuable employee can create the tools if need be.

- Standards are:
  - Standard fabrication and installation details/drawings such as shoe design and base ell supports.

  - o Drawing standards such as layering, text heights, and drawing symbols.
  - o Charts such as line spacing within racks.
- Specifications are: piping classes, equipment spacing requirements, egress and ingress requirements (walkways, platforms and ladders).
- Procedures are: drawing reviews, model reviews, checking, and as-builting.

Below are some brief explanations of standards, specifications, and procedures, and their most likely sources. There are no guarantees, so you will have to investigate each in turn. As we progress into further chapters we will highlight these in more detail, discuss the importance of decision making at an early stage, and discuss the links between the topics. Once you have investigated, assembled, and made all your decisions, you are ready to go, and you have set yourself on a path towards a successful piping execution. By the time you have completed your initial set-up you will have a greater understanding of the project, the expectations, and how you will achieve those expectations. Knowing the reasons behind all of the decisions you have made or helped to make will put you in a position to recognize when things are going wrong, and will aid in correcting them.

1.2 Standards

To determine whether the standards to be used are going to come from your own company or your client, you must consult with your project management team and the client.

Standards include:

- Standard drawings.
- Charts.
- Drawing templates and drawing standards.
- Drawing numbering.
- 3-D model numbering.
- Material commodity codes.

1.2.1 Standard Drawings

Standard drawings are typical fabrication and installation details of commonly encountered items, and are assigned a tag number for easy reference on the piping arrangements and isometrics. The use of a standard avoids detailing the same thing time after time. Commonly, standard drawings are:

- Shoes.
- Anchors: fixed and directional.
- Guides.
- Base ell supports.
- Dummy legs.
- Trunnions.
- Field supports.
- Reinforcing pads.
- Slide plates.
- Tracing details.
- Insulation details.
- Instrument connection details.
- Orifice tap orientations.
- Block and bleed details.
- Vents and drains.

Standards are designed to cover a range of Nominal Pipe Size (NPS) sizes so that one standard may be used on a number of pipe sizes. For instance, all companies will have shoe designs that will cover a range similar to the ones below:

- NPS 6 and under.
- NPS 8 to NPS 12.
- NPS 14 to NPS 18.
- NPS 20 to NPS 24.

You will find that companies mercilessly plagiarise from each other, and most likely you will recognize standards that you have used before as you move from one company to another. You may even see a standard that you created or helped to create being used by another company.

Examples of standards are shown in Figures 1–1, 1–2, 1–3, and 1–4.

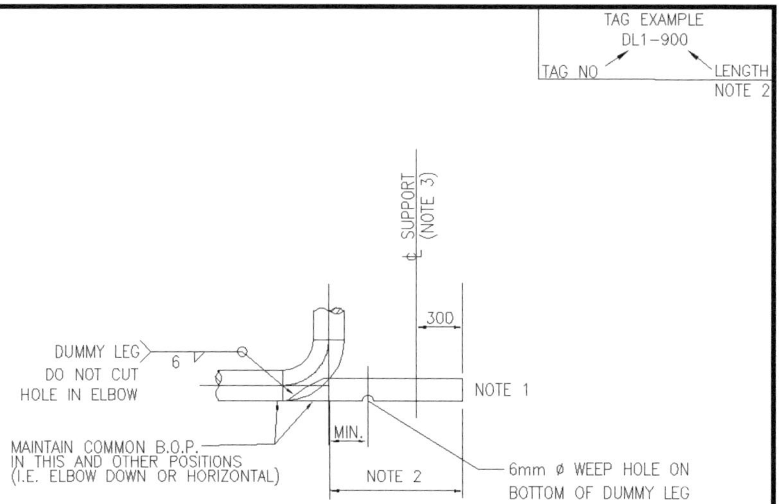

PIPE NPS	DUMMY LEG NPS
2	1 1/2 SCH. 80
3	2 SCH. 80
4	3 STD. WT.
6	4 STD. WT.
8	6 STD. WT.
10	6 STD. WT.
12	8 STD. WT.
14	8 STD. WT.
16	10 STD. WT.
18	10 STD. WT.
20	12 STD. WT.
24	12 STD. WT.

NOTES:

1. DUMMY SUPPORT MATERIAL TO MATCH PIPING CLASS.
2. CENTRELINE OF ELBOW TO END OF DUMMY LEG LENGTH TO BE NOTED ON THE ISOMETRIC.
3. SHOE TO BE ADDED WHERE REQUIRED.

Figure 1–1 *Dummy Leg Standard.*

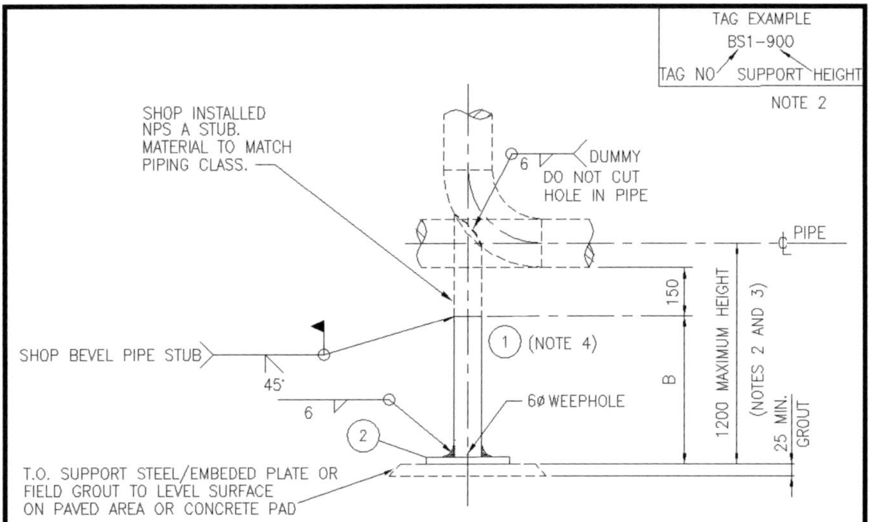

PIPE NPS	NPS A	SHOP FAB B DIM.	C SQ.	T
2	1 1/2 SCH. 80	1100	100	10
3 AND 4	2 SCH. 80	1100	150	10
6 AND 8	4 STD. WT.	1050	200	10
10 AND 12	6 STD. WT.	1000	250	12
14 AND 16	8 STD. WT.	950	300	12
18 AND 20	10 STD. WT.	900	350	16
24	12 STD. WT.	850	400	16
26 AND 36	18 STD. WT.	750	500	20

ITEM No.	DESCRIPTION	MATERIAL	QUANTITY
1	NPS A x B HIGH	A-106 B	1
2	BASE ℞ T THK. x C x C	G40.21M GR. 300W	1

NOTES:

1. ITEM 1 SUPPORT TO BE CENTERED ON ITEM 2 BASE PLATE.
2. CENTRELINE OF ELBOW/PIPE TO UNDERSIDE OF BASE PLATE TO BE NOTED ON THE ISOMETRIC.
3. HEIGHTS GREATER THAN 1200 ARE TO BE DISCUSSED WITH THE STRESS ENGINEER.
4. FIELD TRIM 'B' DIMENSION TO SUIT FINAL PIPING CENTRELINE ELEVATION.

Figure 1–2 *Base Support Standard.*

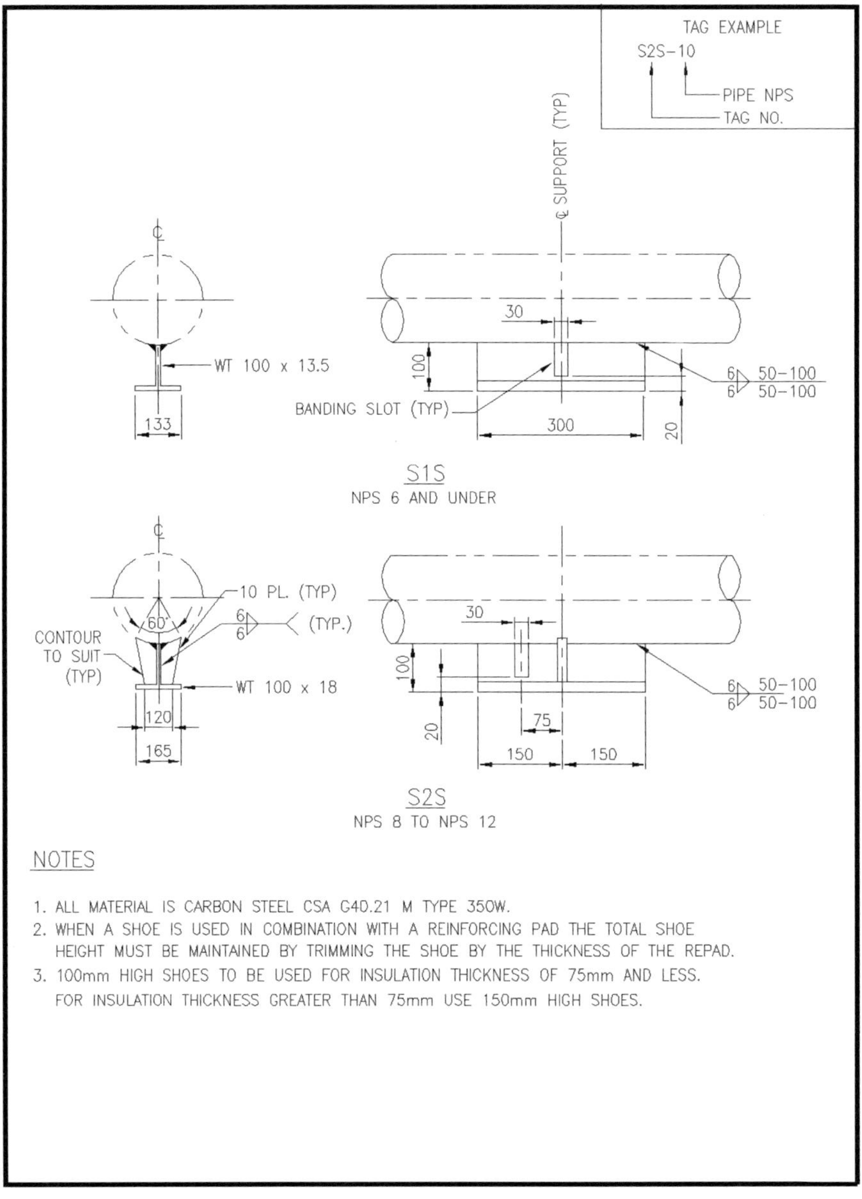

Figure 1–3 *Shoe Standard.*

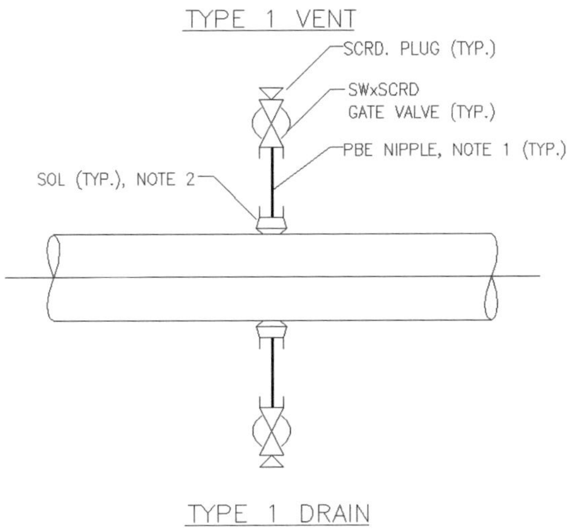

NOTES:

1. USE 100mm LONG NIPPLE FOR 75mm AND LESS INSULATION THICKNESSES. FOR GREATER THAN 75mm INSULATION THICKNESSES USE 150mm LONG NIPPLE.
2. USE AN ELBOLET WHEN VENT IS LOCATED ON AN ELBOW.
3. ALL MATERIALS AND RATINGS TO MATCH THE PIPING CLASS OF THE PIPING SYSTEM.
4. SEE THE ISOMETRIC DRAWINGS FOR VENT AND DRAIN LOCATIONS.
5. U.N.O. ON THE P&IDS ALL PROCESS VENTS AND DRAINS ARE NPS 3/4. PROCESS ENGINEER MAY INCREASE SIZE AS DEEMED NECESSARY.
6. ALL HYDROSTATIC VENTS AND DRAINS ARE AS FOLLOWS:
   ALL NPS LINE SIZES – VENTS ARE NPS 3/4.
   LINES NPS 8 AND BELOW – DRAINS ARE NPS 3/4.
   LINES NPS 10 AND NPS 12 – DRAINS ARE NPS 1.
   LINES NPS 14 AND ABOVE – DRAINS ARE NPS 1 1/2.

Figure 1–4 *Single Block Vent and Drain Standard.*

1.2.2 Charts

There are three charts that are the most important to the piping designers and must be in place:

- Branch connection (can vary by piping class and may be included within the piping classes).
- Line spacing.
- Line spanning.

Examples of these are shown in Tables 1–1a, 1–1b, 1–2, and 1–3.

While line spacing charts and branch connection charts are fairly straightforward, care must be taken with line spanning charts. Line spanning charts will provide layout guidance, but many load factors may affect the posted spans. Final pipe spans must be confirmed during stress analysis.

Table 1–1a Branch Connections Instrument and Utility Air

		Branch Pipe Size (NPS)						
		1/2	3/4	1	1 1/2	2	3	4
Run Pipe Size (NPS)	1/2	T						
	3/4	RT	T					
	1	RT	RT	T				
	1 1/2	T&S	RT	RT	T			
	2	T&S	T&S	RT	RT	T		
	3	T&S	T&S	T&S	RT	RT	T	
	4	T&S	T&S	T&S	T&S	RT	RT	T

T = Straight Tee with threaded ends
RT = Reducing Tee with threaded ends
T&S = Straight Tee and Swage with threaded ends

Table 1–1b Branch Connections Process Lines

Run Pipe Size (NPS)	Branch Pipe Size (NPS)															
	1/2	3/4	1	1 1/2	2	3	4	6	8	10	12	14	16	18	20	24
1/2	T															
3/4	RT	T														
1	RT	RT	T													
1 1/2	T&S	RT	RT	T												
2	Olet	Olet	Olet	T&S	T											
3	Olet	Olet	Olet	Olet	RT	T										
4	Olet	Olet	Olet	Olet	WOL	RT	T									
6	Olet	Olet	Olet	Olet	WOL	RT	RT	T								
8	Olet	Olet	Olet	Olet	WOL	WOL	RT	RT	T							
10	Olet	Olet	Olet	Olet	WOL	WOL	RT	RT	RT	T						
12	Olet	Olet	Olet	Olet	WOL	WOL	WOL	RT	RT	RT	T					
14	Olet	Olet	Olet	Olet	WOL	WOL	WOL	RT	RT	RT	RT	T				
16	Olet	Olet	Olet	Olet	WOL	WOL	WOL	RT	RT	RT	RT	RT	T			
18	Olet	Olet	Olet	Olet	WOL	WOL	WOL	WOL	RT	RT	RT	RT	RT	T		
20	Olet	Olet	Olet	Olet	WOL	WOL	WOL	WOL	RT	RT	RT	RT	RT	RT	T	
24	Olet	Olet	Olet	Olet	WOL	WOL	WOL	WOL	WOL	RT	RT	RT	RT	RT	RT	T

Olet = socketweld or threaded end and rating as defined in the applicable piping class
WOL = bevelled end and schedule as defined in the applicable piping class
T = Straight Tee with ends and schedule or rating as defined in the applicable piping class
RT = Reducing Tee with ends and schedule or rating as defined in the applicable piping class
T&S = Straight Tee and Swage with ends and schedule as defined in the applicable piping class
Note: Reducing Tees may be used in place of Olet/WOL if available.

Table 1–2 Line Spacing Chart

RATING	150#													
SIZE	1"	1 1/2"	2"	3"	4"	6"	8"	10"	12"	14"	16"	18"	20"	24"
150# 1"	115													
1 1/2"	130	130												
2"	135	145	155											
3"	155	165	165	180										
4"	180	180	180	205	205									
6"	205	205	205	230	230	255								
8"	230	230	230	255	255	305	335							
10"	255	255	280	280	305	335	360	385						
12"	305	305	305	335	335	360	385	410	435					
14"	335	335	335	360	360	385	410	435	460	485				
16"	360	360	360	385	385	435	460	485	510	510	535			
18"	375	385	385	410	410	435	460	485	535	535	560	585		
20"	410	410	410	435	435	460	485	535	560	560	585	610	640	
24"	460	460	485	485	510	535	560	585	610	610	640	665	690	740

RATING	300#													
SIZE	1"	1 1/2"	2"	3"	4"	6"	8"	10"	12"	14"	16"	18"	20"	24"
300# 1"	130													
1 1/2"	130	130												
2"	130	155	155											
3"	155	180	180	180										
4"	180	180	205	205	230									
6"	205	230	230	230	255	280								
8"	255	255	255	280	280	305	335							
10"	280	280	280	305	305	360	385	410						
12"	305	335	335	335	360	385	410	435	450					
14"	360	360	360	385	385	410	435	460	490	510				
16"	385	385	385	410	410	460	490	510	535	535	560			
18"	410	410	435	435	460	490	510	535	560	560	585	610		
20"	435	460	460	460	490	510	535	560	585	610	640	665	690	
24"	510	510	535	535	560	585	610	640	665	665	690	720	740	790

RATING	150#													
SIZE	1"	1 1/2"	2"	3"	4"	6"	8"	10"	12"	14"	16"	18"	20"	24"
300# 1"	130	130	130	155	180	205	230	255	305	330	360	385	410	460
1 1/2"	130	130	155	155	180	205	230	255	305	330	360	385	410	460
2"	130	155	155	155	180	205	230	280	305	335	360	385	410	485
3"	155	180	180	180	205	230	255	280	335	360	385	410	435	485
4"	180	180	205	205	230	255	280	305	335	360	385	410	435	510
6"	205	230	230	230	255	280	305	335	360	385	435	435	460	535
8"	255	255	255	280	280	305	335	360	385	410	460	460	485	560
10"	280	280	280	305	305	360	385	410	435	435	485	485	535	585
12"	305	335	335	335	360	385	410	435	460	485	510	535	560	610
14"	360	360	360	385	385	410	435	460	485	510	535	560	585	640
16"	385	385	385	410	410	460	485	510	535	535	560	585	610	665
18"	410	410	435	435	460	485	510	535	560	560	585	610	640	690
20"	435	460	460	460	485	510	535	560	585	610	640	665	690	740
24"	510	510	535	535	560	585	610	640	665	665	690	715	740	790

RATING	300#													
SIZE	1"	1 1/2"	2"	3"	4"	6"	8"	10"	12"	14"	16"	18"	20"	24"
600# 1"	130	130	130	155	180	205	255	280	305	360	385	410	435	510
1 1/2"	130	130	155	180	180	230	255	280	335	360	385	410	460	510
2"	130	155	155	180	205	230	255	280	335	360	385	435	460	535
3"	155	180	180	180	205	230	280	305	335	385	410	435	460	535
4"	205	205	205	230	230	255	280	305	360	385	410	460	490	560
6"	230	230	255	255	280	305	335	360	385	410	460	490	510	585
8"	255	280	280	280	305	335	360	385	410	435	490	510	535	610
10"	305	305	335	335	360	385	410	435	460	460	510	535	560	640
12"	335	335	360	360	385	410	435	460	490	485	535	560	585	665
14"	360	360	385	385	410	435	460	485	510	510	535	560	610	665
16"	410	410	410	435	435	460	485	510	535	560	585	610	640	690
18"	435	435	435	460	460	485	510	535	560	585	610	640	665	715
20"	460	460	485	485	510	535	560	585	610	610	640	665	690	740
24"	535	535	535	560	560	585	610	640	665	690	715	740	765	815

RATING	150#													
SIZE	1"	1 1/2"	2"	3"	4"	6"	8"	10"	12"	14"	16"	18"	20"	24"
600# 1"	130	130	130	155	180	205	230	255	305	335	360	385	410	460
1 1/2"	130	130	155	155	180	205	230	255	305	335	360	385	410	460
2"	130	155	155	155	180	205	230	280	305	335	360	385	410	490
3"	155	180	180	180	205	230	255	280	335	360	390	410	435	490
4"	205	205	205	230	230	255	280	305	335	360	390	410	435	510
6"	230	230	255	255	280	305	335	360	390	385	435	435	460	535
8"	255	280	280	280	305	335	360	385	410	435	460	485	510	560
10"	305	305	335	335	360	385	410	435	460	460	485	510	535	585
12"	335	335	360	360	385	410	435	460	490	485	510	535	560	610
14"	360	360	385	385	410	435	460	485	510	510	535	560	585	640
16"	410	410	410	435	435	460	485	510	535	560	585	610	640	690
18"	435	435	435	460	460	490	510	535	560	585	610	640	665	715
20"	460	460	490	490	485	535	560	585	610	610	640	665	690	740
24"	535	535	535	560	560	585	610	640	665	690	720	740	765	815

RATING	600#													
SIZE	1"	1 1/2"	2"	3"	4"	6"	8"	10"	12"	14"	16"	18"	20"	24"
600# 1"	130													
1 1/2"	130	130												
2"	130	155	155											
3"	155	180	180	180										
4"	205	205	205	230	230									
6"	230	230	255	255	280	305								
8"	255	280	280	280	305	335	360							
10"	305	305	335	335	360	410	410	435						
12"	335	335	360	360	385	410	435	460	485					
14"	360	360	385	385	410	435	460	485	510	510				
16"	410	410	410	435	435	460	485	510	535	560	585			
18"	435	435	435	460	460	485	510	535	560	585	610	640		
20"	460	460	485	485	510	535	560	585	610	610	640	665	690	
24"	535	535	535	560	560	585	610	640	665	690	715	740	765	815

	LINES WITHOUT FLANGES													
SIZE	1"	1 1/2"	2"	3"	4"	6"	8"	10"	12"	14"	16"	18"	20"	24"
LINES WITHOUT FLANGES 1"	85													
1 1/2"	95	100												
2"	100	105	115											
3"	115	120	125	140										
4"	130	135	140	155	165									
6"	155	160	165	180	195	220								
8"	180	185	190	205	220	245	270							
10"	210	215	220	235	245	275	300	325						
12"	235	240	245	260	270	300	325	350	375					
14"	250	255	260	275	285	315	340	365	390	410				
16"	275	280	285	300	315	340	365	390	420	435	460			
18"	300	305	310	325	340	365	390	420	445	460	485	510		
20"	325	330	335	350	365	390	415	445	470	485	510	535	560	
24"	375	380	385	400	415	440	465	495	520	535	560	585	610	660

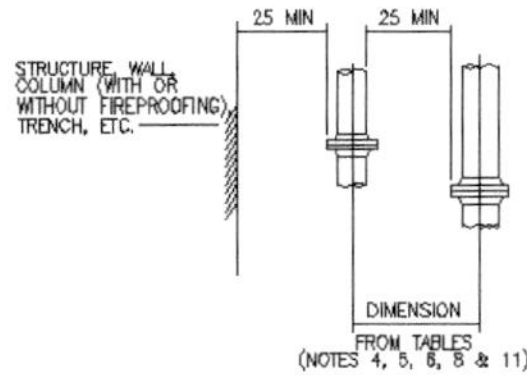

LINES WITH FLANGES

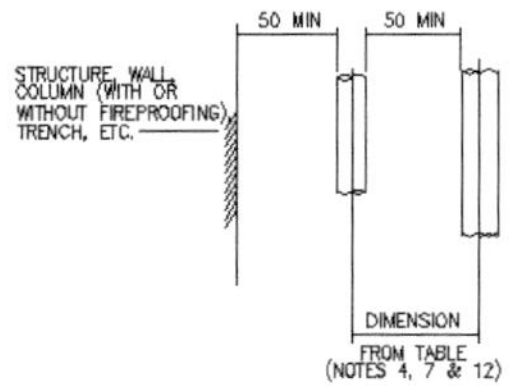

LINES WITHOUT FLANGES

NOTES:

TO OBTAIN A PRELIMINARY ESTIMATE OF THE PIPEWAY WIDTH REQUIRED FOR A SELECTION OF LINES WITHOUT FLANGES, IN THE SIZE RANGE NPS 2 THRU NPS 8, EITHER OF THE FOLLOWING FACTORS MAY BE USED:

1. IF ALL PIPE SIZES ARE KNOWN, ADD NOMINAL SIZES IN INCHES TOGETHER & MULTIPLY BY 0.34 TO ESTIMATE THE WIDTH IN FEET. (THIS IS THE PREFERRED METHOD.)
2. IF ONLY THE NUMBER OF LINES IS KNOWN, MULTIPLY THIS NUMBER BY 1.43 TO ESTIMATE THE WIDTH IN FEET.

EITHER FACTOR GIVES A PIPEWAY WIDTH WHICH INCLUDES INSULATION FOR 25% OF LINES, ALLOWS 20% OF THE WIDTH FOR THE ADDITION & RE-SIZING OF LINES, & ALLOCATES A FURTHER 20% OF THE WIDTH FOR FUTURE PIPING.

3. LINE SPACING DIMENSIONS LISTED ARE BASED ON 25mm FLANGE TO PIPE CLEARANCE (ROUNDED OFF TO EVEN DIMENSIONS) WITH FLANGES ON LARGER LINE & ALL LINES BEING ON A 100mm HIGH SHOE.
4. WHERE AN INSULATED & BARE LINE RUN ADJACENT, THE INSULATION THICKNESS IS TO BE ADDED TO THE SPACING DIMENSION.
5. WHERE TWO INSULATED LINES WITH UNINSULATED FLANGES RUN ADJACENT, THE LARGER OF THE TWO INSULATION THICKNESSES IS TO BE ADDED TO THE SPACING DIMENSION.
6. WHERE TWO INSULATED LINES RUN ADJACENT & EITHER ONE OR BOTH LINES HAVE INSULATED FLANGES, TWICE THE LARGER OF THE TWO INSULATION THICKNESSES IS TO BE ADDED TO THE SPACING DIMENSION.
7. WHERE TWO INSULATED LINES WITHOUT FLANGES RUN ADJAC THE COMBINED INSULATION THICKNESSES ARE TO BE ADDED THE SPACING DIMENSION.
8. REVIEW THE 25mm GAP BETWEEN ADJACENT LINES AT CORNERS, EXPANSION LOOPS, ETC. TO ENSURE ADEQUATE SPACING FOR THERMAL MOVEMENT.
9. INCREASED SPACING MAY BE REQUIRED WHERE ORIFICE FLANGES ARE REQUIRED.
10. INCREASED SPACING MAY BE REQUIRED WHERE PIPE CLAMPS ARE USED. (e.g. ANTI-VIBRATION SERVICE, FRP PIPE SUPPORTS, ETC.)
11. FOR LINES WITH CLASS 900 AND ABOVE FLANGES CALCULA THE REQUIRED SPACING.
12. DESIGNER MAY USE ONLY WHEN CERTAIN THAT THE POSSIBI OF ADDING FLANGES INLINE IN THE FUTURE DOES NOT EXIS NOT RECOMMENDED FOR THE SPACING OF LINES IN PIPERA

Table 1–3 Line Spanning Chart

PIPE SUPPORT SPACING								
PIPE SIZE (NPS)	PIPE O.D. (mm)	PIPE SCHEDULE (NOTES 3,4,5)	WALL THICKNESS (mm)	WEIGHT OF WATER–FILLED PIPE SPAN (KILOGRAMS)	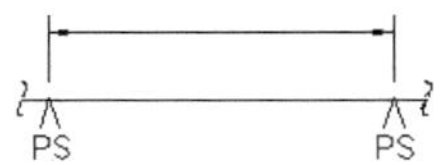		 (NOTE 7)	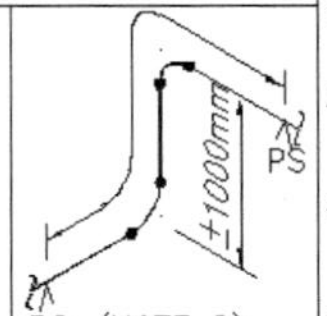 (NOTE 8)
					ALLOWABLE SPAN (m)	DEFLECTION (mm)	FLAT TURN (m)	ELEVATED TURN (m)
3/4	26.7	160	5.537	–	3.6	15	2.3	2.7
		80	3.912	–	3.0	10	2.0	2.3
1	33.4	160	6.350	21.77	4.8	6	3.1	3.6
		80	4.547	18.14	4.0	13	2.6	3.0
1 1/2	48.3	160	7.137	47.63	5.9	6	3.8	4.4
		80	5.080	38.56	5.2	19	3.4	3.9
2	60.3	80	5.537	61.69	6.1	25	4.0	4.6
		40	3.912	48.53	5.7 (NOTE 5)	15	3.7	4.3
3	88.9	80	7.620	155.13	7.9	6	5.2	6.0
		40	5.486	123.84	7.2	15	4.7	5.4
4	114.3	80	8.560	264.89	8.9	6	5.8	6.7
		40	6.020	207.75	8.1	15	5.3	6.1
6	168.3	80	10.973	633.21	10.7	6	6.9	8.0
		40	7.112	469.47	9.8	15	6.4	7.4
8	219.1	80	12.700	1128.99	12.0	6	7.8	9.0
		40	8.179	832.80	11.2	5	7.3	8.4
		20	6.350	703.52	10.5	4	6.8	7.9
10	273.1	80	15.062	1892.39	13.3	6	8.7	10.0
		40	9.271	1354.88	12.2	5	7.9	9.2
		20	6.350	1054.15	11.1	4	7.2	8.3
12	323.9	80	17.450	2853.09	14.5	6	9.4	10.8
		STD	9.525	1989.47	13.3	15	8.6	10.0
		20	6.350	1451.04	11.5	4	7.5	8.6
14	355.6	80	19.050	3575.67	15.2	6	9.9	11.4
		STD	9.525	2477.97	13.8	15	9.0	10.3
		20	7.925	1989.00	12.5	4	8.1	9.4
16	406.4	80	21.412	4959.58	16.1	6	10.5	12.1
		STD	9.525	3465.45	14.5	15	9.4	10.9
		20	7.925	2536.94	12.9	4	8.4	9.7
18	457	80	23.800	6597.50	17.1	6	11.1	12.8
		STD	9.525	4667.01	15.1	15	9.8	11.3
		20	7.925	3167.89	13.2	3	8.6	9.9
20	508	80	26.187	8521.19	18.0	5	11.7	13.5
		STD	9.525	4333.17	14.2 (NOTE 6)	3	9.2	10.7
24	610	80	30.937	13308.85	19.6	5	12.8	14.7
		STD	9.525	6094.92	14.7 (NOTE 6)	3	9.5	11.0

NOTES:

1. SPANS ARE BASED ON INSULATED CARBON STEEL PIPE (A106–B) FILLED WITH WATER FROM –29°C TO 260°C.
2. XS PIPE AND SCH. 80 HAVE THE SAME WT IN ALL NPS SIZES THRU NPS 8. FROM NPS 10 THRU NPS 24 XS PIPE HAS A WALL THICKNESS 12.7mm.
3. STD PIPE AND SCH. 40 HAVE THE SAME WT IN ALL NPS SIZES THRU NPS 10. FROM NPS 12 THRU NPS 24 STD PIPE HAS A WALL THICKNESS OF 9.525mm
4. STD PIPE AND SCH. 20 HAVE THE SAME WT IN NPS 20 NPS AND NPS 24.
5. ALLOWABLE SPAN MAY BE INCREASED TO 6.0m FOR NPS 2 SCH. 40 ON STRAIGHT RUNS ALONG PIPEWAYS.
6. NPS 20 AND NPS 24 MAY REQUIRE PADS OR SADDLES.
7. 65% OF ALLOWABLE SPAN
8. 75% OF ALLOWABLE SPAN.
9. SEE STRESS GROUP IF VALUES SHOWN ARE EXCEEDED.

1.2.3 Drawing Templates and Drawing Standards

Drawing templates are required for the drawings that are to be created for the project. There are four common drawing templates for four plot sizes:

- ANSI paper sizes used in the USA and Canada.
  - A size - 8½" × 11" for standards.
  - B size - 11" × 17" for isometrics.
  - D size - 22" × 34" for Process Flow Diagrams (PFDs), Piping and Instrumentation Diagrams (P&IDs), and piping arrangements.
  - E size - 34" × 44" for plot plans, equipment location plans, and key plans.
- ISO A series paper sizes used in the rest of the world.
  - A4 - 210mm × 297mm for standards.
  - A3 - 297mm × 420mm for isometrics.
  - A1 - 594mm × 841mm for PFDs, P&IDs, and piping arrangements.
  - A0 - 841mm × 1189mm for plot plans, equipment location plans, and key plans.

ANSI C (17" × 22") and ISO A2 (420mm × 594mm) paper sizes are usually reserved for reduced size plots of ANSI D and E, and ISO A1 and A0. This is due to being a convenient handling size while retaining a large enough drawing size for clarity and mark-ups.

It is most likely that the client will have drawing templates for three of the four drawing sizes, ANSI A, D, and E, or ISO A4, A1, and A0, that they will require you to use on their project. For the isometric template, ANSI B or ISO A3, the client will most often defer to the engineering company. Clients take isometrics back for future reference, but they are considered fabrication documents and are not issued out to other engineering companies to be revised on future projects.

The need to insist on a client titleblock and client drawing numbering system does not apply to the isometrics like it does to the other drawings, and engineering companies have commonly produced isometrics on their own drawing template.

Unlike 2-D projects where isometrics are drafted by following the checked piping arrangements and are reviewed but not stamped by an engineer, engineers are stamping the isometrics produced on 3-D projects. This is because the accuracy of the automatic isometrics are directly dependent on the output from the 3-D model databases, the decisions and inputs for which are made by the designer.

Drawing templates have predetermined drawing standards as part of their set-up, so that when a drawing is plotted, the text heights, line weights, etc., will be to the correct dimensions.

Drawing standards include:

- Titleblock with company logo.
- Text heights.
- Layering system.
- Line weights.
- Dimension styles.

Other forms of drawing standards that you will require are drawing symbols legend sheets and drafting abbreviations. Primarily, the legend sheets you will require are going to be the PFD and P&ID legend sheets from the client that contain all the approved symbols to be used on the drawings. Drafting abbreviations come from various industry organizations, e.g., ANSI, ASME and ISA. The most commonly used ones are often summarized as a company standard. See Appendix A for an example of drafting abbreviations used on piping drawings.

1.2.4 Drawing Numbering

Clients have drawing numbering requirements. Each client mandates the drawing numbering to be used. File numbering, which ideally should equate to the drawing number, is also mandated in order for the client to be able to accept the drawing files back into the document management system and be able to retrieve them when required. While numbering systems vary by company, most have a hierarchy numbering system using abbreviated identifiers along the following lines:

- Area of operation.
- Facility.
- Discipline. E.g., mechanical, piping.
- Type of drawing. E.g., piping arrangement.
- Three or four digit sequential drawing number.
- Two or three digit sequential sheet number.

1.2.5 3-D Model Numbering

What is the model numbering convention to be used for the project? Again, this may be a client or engineering company standard, but it must be decided right at the beginning of the project. If you are directed to use a company standard, make sure that your client is in agreement. These models will later have to be closed-out and renumbering/renaming can cause considerable work that is likely not included in the budget. As with drawing file numbering, models require a file numbering system that is approved by the client. In the 3-D world of today's integrated design, where all disciplines are referencing the other disciplines' models, and working more-or-less in real time, you and your designers must intimately know the model numbering system. This is in order to correctly name the piping models and identify other disciplines' models that are required to be referenced.

One way that I have seen model numbering done is to follow the Construction Work Package (CWP) numbering. This makes sense because assembling information is not just the domain of the designers. Often the CAD support group or the material control group will be requested to generate reports of a particular CWP, or a group of CWPs, and a common numbering system for the models and the CWPs will help to locate and compile the information. Common reports are as follows:

- Material Take Off (MTO), either bulk or item specific for:
  - Pipe and fittings.
  - Insulation.
  - Valves.
  - Shop and field material split.
- Weld count and diameter inches of welding.
- Weights of materials.

As all of the reports are generated from the databases that are built as the models are developed, having a direct correlation to the CWP number makes life easier for the downstream people, such as material controllers and purchasers.

1.2.6 Material Commodity Codes

Material commodity codes are a component identification numbering system used for the identification, ordering, and tracking of materials. Originating in the 3-D model material library database, these numbers appear on the Material Take Off (MTO) reports and the isometric Bills of Material (BOM) and in the documentation used by the purchasers, suppliers, fabricators, and warehousing. These codes are an alpha-numeric string which uniquely identifies a component. Industry standards do not exist for material commodity coding of piping components, so companies have to develop their own key element identifiers, e.g., possibly an 'F' for flange.

So why should this be a concern for the piping lead? If you are using your company piping classes and the material commodity coding is in place, then for the most part you and your designers will have no interest in material commodity codes other than curiosity. However, if material commodity codes are to be used on the project, and development is required because your company does not currently have a commodity coding system, or the client wants you to use a different commodity coding system, then you will need to take into consideration the time involved for the development and implementation by the material control group and IT. Development and implementation can impact your ability to start 3-D modeling.

As not all projects use commodity codes, you will have to ascertain whether they are to be used on your project or not. If they are, are they to be to your company's standards or the client's? Will the adoption of a commodity code numbering system cause you any delays?

1.3 Specifications

Many clients and engineering companies will have a set of company specifications for each discipline built on code, safety and insurance requirements, and preferred engineering practices. Most specifications are engineering related, but some are directly related to the layout of the plant. As has already been said, these are related to walkways, platform and ladder requirements, egress and ingress, and equipment spacing, but they often also include other information of importance to the piping designers such as pump and exchanger piping layouts, and transportation requirements for modules and spools. As the lead, you must review these and select all the pertinent specifications for your team. While it may be that all the specifications are on the company network, in reality most people don't take the time to sort through them. You must do this as a part of your project set-up: copy the specifications into a piping specific directory and create a set in hard copy format for reference.

The most obvious and essential specifications to be secured are the piping classes.

1.3.1 Piping Classes

The piping classes are one of the most important specifications for the piping designer. These are developed by the piping engineer, and most often have been applied on numerous projects, sometimes for many years. Piping classes are developed around the applicable piping code and list components manufactured to the standards listed within the code. This allows component use with no further investigation and avoids calculations and material selections being repeated time and time again for the same application. The piping classes list the following:

- ASME code (B31.1 in power plants and B31.3 in process plants).
- Service/commodities.
- Flange rating.
- Corrosion allowance.
- Temperature range.
- Pressure limits.
- Non-Destructive Examination (NDE).
- Heat treatment.
- NPS range, and pipe schedule wall thickness (WT) or calc. wall.
- End preparation.
- Valves.
- Listed standards/components and materials accepted by the code for:
    - Pipe.
    - Fittings.
    - Flanges.
    - Orifice flanges.
    - Unions.
    - Plugs.
    - Nipples.
    - Spectacle blinds, spades, and spacers.

- Olets.
- Gaskets.
- Bolting.

An example of a piping class is Table 1–4.

Table 1–4 Example Piping Class

PIPING CLASS AAA – CLASS 150 RATING

SERVICE: Sweet Hydrocarbons, Fuel Gas.

PRESSURE LIMIT @ TEMPERATURE:

Temp. °F (°C)	–20 to100 (–29 to 38)	200 (93)	300 (149)	400 (204)	500 (260)
MAWP, psig (kPag)	285 (1965)	260 (1793)	230 (1586)	200 (1379)	170 (1172)

ASME B31.3 LATEST EDITION

Required Corrosion Allowance:	1.6 mm/$^{1}/_{16}$"
Material Group:	P1 Groups 1 and 2; Carbon Steel
Inspection:	100% Visual Inspection, 100% RT of circumference on 10% of butt welds per welder/welding operator, progressive production basis.
Heat Treatment:	On welds >19mm/¾" in thickness (ASME B31.3 Table 331.1.1)
Maximum Hardness:	200 Brinell Number

Table 1–4 Example Piping Class (Continued)

Pipe, Flanges and Fittings:

ITEM/CODE	NPS SIZE	RATING	CONNECTION	MATERIAL STANDARD
PIPE: ASME B36.10M	¾ – 1½ ¾ – 1½ 2 – 24	Sch 160 Sch 80 Std.	Thr'd End (TE) Plain End (PE) Bevelled End (BE)	A106 Gr. B A106 Gr. B A106 Gr. B
NIPPLES: ASME B36.10M	½ – 1½	XXS	TE, PE	A106 Gr. B
SWAGES: ASME B16.9	¾ – 1½	XXS	TE, PE	A234 Gr. WPB
FLANGES: ASME B16.5	¾ – 1½ 2 – 24	Class 150 RF Class 150 RF (Std)	Thr'd, SW WN	A105N A105N
ORIFICE FLANGES: ASME B16.36	1 – 1½ 2 – 24	Class 300 RF Class 300 RF (Std)	Thr'd, SO WN	A105N A105N
INSTRUMEN T FLANGES: ASME B16.5	¾ – 1½ 2 – 24	Class 300 RF Class 300 RF (Std)	Thr'd, SW WN	A105N A105N
FITTINGS: ASME B16.11 ASME B16.9	¾ – 1½ 2 – 24	3000# Std	Thr'd, SW BW	A105N A234 Gr. WPB
UNIONS: ASME B16.11	¾ – 1½	3000#	Thr'd, SW	A105N
OLETS: ASME B16.11 ASME B16.9	¾ – 1½ 2 – 24	3000# Std	TOL, SOL WOL	A105N A105N
BLINDS: ASME B16.5	¾ – 12 14 – 24	Class 150 RF Class 150 RF	Spectacle Blinds Spades/Spacers	A516 Gr. 70N A516 Gr. 70N
PLUGS: ASME B16.14	½ – 1½	3000# Solid Hex Head	Thr'd	A105N
BOLTING: ASME B18.2.1			Studs Hex Nuts	A193 Gr. B7 A194 Gr. 2H
GASKETS: ASME B16.20	¾ – 24	Class 150 RF	Spiral Wound 3.2 mm	316 SS, non-asbestos, inner ring

Table 1–4 Example Piping Class (Continued)

Valve Specifications:
(ASME B16.5, B16.10, B16.11, B16.25, B16.34, API 598)

VALVE TYPE	NPS SIZE	RATING	CONNECTION	VALVE CODE (Varies by company)
GATE: API 600, 602	½ – 1 ½ 2 – 12 14 – 24	Class 800 Class 800 Class 800 Class 150 Class 150, C/W Gear Operator	Thr'd SW SW × Thr'd RF RF	Describes valve design, e.g., bolted bonnet, flexible wedge, regular port, OS & Y.
GLOBE: API 600	¾ – 1 ½ 2 – 4 6 – 24	Class 800 Class 800 Class 800 Class 150 Use Gate Valve	Thr'd SW SW × Thr'd RF RF	Describes valve design, e.g., bolted bonnet, stem guided, OS & Y.
BALL: API 608	¾ – 1 ½ 2 – 6 8 – 24	Class 800 Class 800 Class 800 Class 150 Class 150, C/W Gear Operator	Thr'd SW SW × Thr'd RF RF	Describes valve design, e.g., split body, floating ball, regular port.
CHECK: API 594, 600	¾ – 1 ½ 2 – 24	Class 800, Lift Class 800, Lift Class 150, Swing	Thr'd SW RF	Describes valve design, e.g., bolted cap, swing disc.
NEEDLE:	¾ – 1	Class 6000, MNPT × FNPT Class 6000, FNPT × FNPT	Thr'd Thr'd	Describes valve design.

Table 1–4 Example Piping Class (Continued)

Valve Materials (ASTM, NACE, Specified Material)

VALVE TYPE	BODY	BONNET	TRIM	BOLTING STUDS/ NUTS
GATE:	A216 WCB or A105	A216 WCB or A105	Stellite – API Trim 8, 13Cr	A193 Gr. B7 A194 Gr. 2H
GLOBE:	A216 WCB or A105	A216 WCB or A105	Stellite – API Trim 8, 13Cr	A193 Gr. B7 A194 Gr. 2H
BALL:	A216 WCB or A105	A216 WCB or A105	PTFE/316SS	A193 Gr. B7 A194 Gr. 2H
CHECK:	A216 WCB or A105	A216 WCB or A105	Stellite – API Trim 8, 13Cr	A193 Gr. B7 A194 Gr. 2H
NEEDLE:	T316SS	Packed T316SS	NACE MR 0175	

NOTES
1. Socket welded piping is preferred over threaded where possible. Use threaded joints at the outlet of vent and drain valves, at outlet of instrument root valves, and to match equipment.
2. Spectacle blinds shall be used up to NPS 12. Spades and spacers shall be used in sizes NPS 14–24.
3. Ball valves have limited maximum temperature (Teflon seats: 200°C, PEEK seats: 220°C).
4. For allowable branch connections refer to Branch Connections Chart.
5. Refer to ASME B16.5 Table 2-2.2 or Table F2-2.2 for Pressure-Temperature Ratings for 316SS Thermowell flanges.

Piping classes have an abbreviated identifier. An example of this is the following three-digit identifier and Table 1–5:

- Flange rating.
- Service.
- Pipe material.

Components manufactured to standards not listed in the code, or not fabricated to a standard, are not listed within the piping classes. These are known as unlisted or specialty items (SP) and they are listed in the specialty item list. These include such items as strainers and expansion joints. In order to be used within the piping system, specialty items must conform to the engineering and testing requirements set out in the code.

Table 1–5 Piping Class Identifiers

Flange Rating	Service	Pipe Material
A – 150	A – Sweet Hydrocarbons	A – Carbon Steel A106B
B – 300	B – Sour Hydrocarbons	B – Carbon Steel A333 Gr. 6
C – 600	C – Process Water	C – CS Galvanized
D – 900	D – Process Steam	D – SS 316
E – 1500	E – Chemicals	E – SS 304
F – 2500	F – Acids	F – FRP
	G – Caustics	G – Plastic PVC
	H – Utilities	H – Polypropylene Lined
		J – PTFE Lined

Piping classes do not take into account out-of-spec items such as a Class 600 flat face flange required to match a pump flange in a Class 300 raised face piping system. Designers will have to request that these out-of-spec items be added to the piping class material database on a case-by-case basis.

There must be an approval procedure for out-of-spec components to be added into the piping classes libraries. A source of frustration for the piping designer is the time delay, and sometimes refusal, to add a component. Likewise, a source of frustration for a material controller is to receive requests to add unnecessary components. As this can result in conflict, it is highly recommended that the piping engineer should approve all material requests; the piping engineer is the custodian of the piping classes, and is the only person that should be authorizing changes and additions.

Calculated wall, or 'calc. wall' as it is often listed in the piping classes, is another subject to be addressed by yourself and the piping engineer as soon as possible. If you start the modeling with the calculated wall it can later become a problem to update the models once the pipe schedule has been determined. This is particularly true when a piping class is being used for several different process commodities and when wall thickness differs in the same NPS range according to the different design conditions.

I strongly suggest that you discuss this sooner rather than later with the piping engineer and the material controller. I also strongly suggest that the calculated wall be based on the worst case design conditions for the piping class. The potential for mistakes in fabrication and material control when different wall thicknesses are applied to the same NPS pipe can outweigh any cost saving that may be realized on the pipe and fittings themselves. It is also possible that a thinner schedule may not be as commercially available as a thicker schedule. Schedule 160, for instance, is usually very available, whereas Schedule 140 usually is not as available.

Finally, the set-up of the piping classes for the use by the designers when creating the 3-D models is a joint exercise between the material control group who creates the database format and the CAD support group who ensures its functionality. Prior to the start of modeling you must have these piping classes checked against the hard copies. The accuracy of all MTO reports and BOM lists on the isometrics rests on the accuracy of the piping class databases and the importance of ensuring the integrity of these databases cannot be overstated.

Once the piping classes have been approved they are frozen for the duration of the project. Changes may happen to the piping classes during the project, but these must follow a deviation process.

You must issue hard copies of the piping classes to your designers. An impression exists that hard copies of the piping classes and branch charts are unnecessary for the piping designers on 3-D projects. This stems from the belief that the choices of available components, including branch fittings, are pre-determined and limited per the piping class databases. However, all your designers should have hard copies for reference because they have a responsibility to understand the root documents and to verify the information when something appears to be incorrect.

1.4 Procedures

Projects revolve around procedures, and without these you cannot execute your project. Procedures are the "Highway Code" that keeps everyone on the same page. However, procedures are often either poorly written or not enforced, which is as bad as having no procedure. Read the procedures thoroughly and understand them, because it will be up to you to enforce them later, and possibly to expand on them.

Engineering company procedures will commonly include:

- Stick files.
- Inter-discipline drawing reviews, both internal (engineered drawings) and vendor.

- Line numbering.
- Stress analysis.
- CAD set-up and support.
- 3-D model reviews.
- Checking.
- Manhour estimating.
- Progress reporting.
- Management of change.

Client procedures have been developed to standardize the drawing and model deliverables. These include:

- As-builting.
- Project close-out.

Be aware that your client will have a document management department who are the custodians of these procedures. Because no procedure is perfect for all situations, this department will be conducive to small deviations and will work with you. On smaller projects you may well have direct access to speak with them, whereas on larger projects you will have to request a deviation through your project management team. Deviations will be documented in the Project Execution Plan (PEP).

The following are brief discussions on all the above mentioned procedures/activities that you, as a piping lead, must be aware of, and that the piping group has a direct or indirect involvement in. In Chapter 2 we will discuss in more detail the procedures below that the piping group manage directly or have a major involvement in.

1.4.1 Stick Files

A stick file is the best tool to use for capturing and managing changes. It is a central depository for all mark-ups from all disciplines, and ensures that all changes are clearly communicated and surprises are avoided.

You will need to identify an area where your master stick files will reside, preferably close by and with reference tables to lay the stick files on. Ideally, you will have room for roller boards. Roller boards, where the drawings are tapped down, work wonderfully to stop people wandering off with your stick file drawings.

1.4.2 Inter-discipline Drawing Reviews

Drawings are produced in every project. Drawings represent the culmination of the design, and a finalization of all discussions and decisions to date. Circulation to the other disciplines that have had input is required in order to establish that the design is as expected by all stakeholders. It is required that both the piping drawings and the vendor drawings be circulated for comment. You are responsible for spearheading the circulation of the piping drawings, and you will require a place to store these circulated drawings afterwards for record keeping. While you are not responsible for the circulation of vendor drawings, you are responsible for maintaining a piping master set for reference and mark-up. For this you will need a filing cabinet with file folders and tabs.

1.4.3 Line Numbering

Line numbering is a standard. The elements that make up a line number and the order of placement are standardized by each client or engineering company. The sequencing of the elements may change between companies, but all line numbers contain the following:

- Piping class.
- NPS.
- Sequential line number.
- Insulation thickness and type.
- Tracing requirements.

Some standards may also include:

- Unit number.
- Commodity abbreviation of the process fluid.

When it comes to the task of assigning line numbers to the P&IDs, a procedure is required. Does a line reduction within a header system constitute the assignment of another line number? Is a pump suction line that splits to a pair of A and B pumps two or three line numbers? You will discover that there are different thoughts on the subject. Line numbering may not be the responsibility of the piping group, and it may fall under the auspices of the process engineers', but you as a piper have a vested interest that requires discussion. The overuse of line numbers can result in piping runs being broken down into a greater number of isometrics.

Whoever does it, and to whatever procedure, line numbering needs to be done sooner rather than later, as modeling cannot begin without line numbers having been assigned.

1.4.4 Stress Analysis

How are you going to interface with the stress group? Which lines will be stressed first? How will you track stress analysis? Where will you store the stress mark-ups? And how will the stress requirements be disseminated to the pipers? These are questions that we will investigate in more detail in Chapter 2. It is a complex matter, and not one to be overlooked. If you don't have a company procedure to fall back on, you will need to develop one.

1.4.5 CAD Set-Up

CAD support is a major contributor to the success of any project. Ensure that the appropriate company and client CAD procedures are going to be used.

- Maintaining databases.
- Plotter and printer set-ups.
- Model back-ups, usually nightly.
- Project close-out.

The above will be of little interest to the designer, and should be happening seamlessly in the background. However, there are project set-up requirements that are of particular importance in order for the designer to do his/her work. But, because there are many ways of doing things, the CAD support group will require direction from you, the piping lead, for the following to be input:

- Work areas.
- Client or company piping classes.
- Client or company piping standards, e.g., shoes, base ell supports, anchors, guides.
- Color coding of pipe, equipment and temporary steel per the client or company standards. For example is pipe to be color coded:
  - By piping classes: Class 150, 300, 600, etc.

  or
  - By commodity: sub-sets of water, steam, oil, gas, etc.

- Link for automatic data extraction from the Line Designation Tables (LDT) into the isometric titleblocks.
- Clash report procedure.

A project directory structure is also required where drawings and models are deposited. This structure should ideally be mirrored by each discipline so that finding models is easy for all. As a side note, designers love to copy into their personal drives and work from there on unofficial models. This practise must be discouraged as it can cause obvious communication problems. There is nothing wrong with copying to do some studies, but the results must be imported into the master model as soon as possible, and all work should primarily only be done in the master model.

Which software and version of that software are you using? Lock it in and do not let anyone tell you that an upgrade during your project is an easy matter. The software and version may be mandated by the client, in which case the decision is off of your shoulders, but if it is a company choice do not change once you start. Changing the current version to an untested version during a project inevitably leads to untold grief and extra, unbudgeted hours. No disrespect to the CAD support group, but even if they tell you that it has been tested, do not let your project be the ultimate test case for the newest version, no matter how much they protest that this will solve many of the problems currently encountered. Testing is nebulous, and their testing of some functionality and file conversion on a small scale will not be representative of the full usage by the designers and conversion of dozens or possibly thousands of models and databases.

Security is another issue to be addressed. Your client may be very interested in this aspect of his/her project, but even if they are not, it is in your own interest to make sure that you are comfortable with the security measures that are in place. Security means access. Who has access to edit the models? Minimally, there should be restrictions on:

- Each discipline. A discipline must not be able to edit the other disciplines' models.
- Stages of design. Freeze models when they are ready for checking. Unfreeze but freeze again after the changes required by checking have been made. Designers are perfectionists and if there are no controls stopping them from doing so, they may go back into their models and make changes during checking and even after it has been Issued For Construction (IFC). While keenness can be admired, unmanaged and unchecked

changes that surface will cause an embarrassment at the least, and can lead to other costly problems.

Several other decisions that you may also be required to have input into with the CAD support manager, office manager, and project manager are:

- The number of CAD stations you will need.
- The types of software and number of licences.
- The space and furniture requirements for common areas, e.g., plotters, printers, stick files, filing cabinets, lay down tables.

When you meet with the office manager be sure to order all of the stamps you will need, i.e., a date stamp, 'RECEIVED,' 'MASTER STICK FILE,' 'WORKING COPY,' 'PIPING COPY,' 'CHECK PRINT,' 'SUPERSEDED,' and 'FOR INFORMATION ONLY.' You could also order stamps for 'STRESS COPY' and 'FOR BID PURPOSES ONLY. NOT TO BE USED FOR CONSTRUCTION,' but these can be added to the drawings as blocks at the time of issue.

Designers require lots of support from the CAD support group. You will require a clear line of communication between the two groups for:

- General questions about CAD execution and CAD commands.
- Copying of models.
- Modeling of specialty items.
- Adding of specialty items and out-of-spec components into the piping specifications and material libraries.
- Retrieving lost data and corrupted files.

Larger engineering companies may well utilize software whereby a request is sent and a ticket number assigned. Smaller companies may well utilize an e-mail request. However this is to be done, make sure that you have a documented procedure that can be distributed to the designers.

1.4.6 3-D Model Reviews

Model reviews are commonly conducted at the 30%, 60% and 90% stages of design completion, and involve buy-in by all stakeholders up to that stage. Definitions are required for each of these stages so that everyone has the same understanding of what is to be accomplished prior to and during the reviews, and the designers stay focused on the parameters to be established leading up to the reviews. The best way for this to be accomplished is to have written documentation.

1.4.7 Checking

It goes without saying that checking is a requirement of any project, but how are you going to go about this? What are you going to check? You need a checking procedure to give guidance and ensure consistency.

1.4.8 Manhour Estimating

Manhour estimating and manpower planning can be quite a daunting task, and there are books written on this subject alone. As a lead you will be required to have input into the piping hours budget and piping schedule. Generally speaking, companies employ schedulers to do nothing but this task, and they will be of valuable assistance in helping you to establish schedules. However, it does not end there. After the piping budget and piping schedule have been established, the task of work allocation—literally the decision on which designers will work in which areas and according to a priority that supports the schedule—rests with the piping lead.

1.4.9 Progress Reporting

Your project management will expect progress reports. Larger engineering companies will have their own procedures to accomplish this. If you are in a smaller company, chances are that you will need to establish a form of reporting in conjunction with the project management. Reporting is essential, not just because periodic payments from the client may be tied to the progress, but because it is necessary to understand where you are in the project, whether or not you are staying within the budgeted hours, and to identify when you are going off track in order to take corrective action.

1.4.10 Management of Change

Change is inevitable in any project, and management of change is crucial to the success of the project. You must manage the trends and scope changes for the piping effort. A trend is an unbudgeted event that increases the number of hours required to complete an activity. For instance, a deviation to a piping class or vendor information that arrives later than planned and causes delays and rework constitutes a trend. A scope change is a modification, deletion, or addition to the original scope that was not budgeted for, such as the addition of a piece of equipment. In order to recognize trends and scope changes, a clear understanding of the project scope and budget is required.

Scope changes are quite easy to spot and usually emanate from the client, whereas trends can be contentious and usually have to emanate from you, the piping lead. Is the change a result of design development or a trend? You're going to find yourself in this debate many times in your career. The simple answer is that there are no simple answers. However, situations such as those below are clues:

- If you have to change your plans and focus due to delayed arrival of information, leading to productivity being lost, this is a trend.
- If you were given information with assurances as to the completeness and accuracy of this information for the level of design, and you have to rework your design because it turned out not to be so, this is a trend.
- If you were required to move ahead with preliminary information into detailed design and later, when firmer information is available, have to rework the design, this is a trend.
- If you or your group has to spend an abnormal amount of time assisting another department or a vendor, this is a trend.

If for whatever reason you find yourself losing productivity and/or reworking a design more than once and are in danger of exceeding the budgeted hours, discuss the situation with your project engineer. A trend may be in order.

You will learn that timing is everything and design developments, such as a line size increase, or scope changes, such as a pump addition, that come during the earlier study stage can be accommodated quite readily without much, if any, schedule impact. But try accommodating the same just before IFC and you will be looking at significant rework, schedule delays and cost impacts.

The surprising thing, considering its importance, is that managing change is often done poorly or not at all. Formally raising, approving/rejecting and documenting these will avoid misunderstandings and wasted hours. I have seen more than one project where the lack of documentation caused significant discord between the engineering company and the client. Clients have a tendency to request changes throughout the project and consider them design development, not trends or scope changes. Engineering companies have a tendency to jump to attention and rush in to accommodate the client. The client believes that his/her requests are going to be accommodated without extra cost or schedule impact while the engineering company assumes that the client realizes otherwise. They are not on the same page, and likely will not be until much later, after further discussions, possibly some hard feelings, and likely after all the changes have already been made.

We must generate change notices to capture the impact that trends or scope changes will have or are having on the piping effort. This documentation allows for a time of assessment to cost and schedule and for a conscious choice to be made on how to proceed. Are the changes necessary? For exmaple, are they safety related, or are they just nice to have? Can something be done about the productivity loss related issues?

Your project management will expect change to be recognized, documented, and submitted for approval. As the eyes and ears of the piping group, the piping lead is expected to keep his finger on the pulse, and not jump the gun and allow unapproved design changes nor keep reworking the same area due to someone else's inability to make up his/her mind. You are expected to see change coming, to anticipate the consequences, and to raise flags ahead of time. Don't put yourself in the unenviable position of trying to explain later why you are going over on budget and schedule. You may end up making changes back to the original design and lose credibility as a lead. Only a junior would use the excuse that someone else told them to do it. Do not forget that you are a senior member of the team with a budget that you are responsible for.

It is important that trends and scope changes be addressed as soon as possible, but you will have very little control over how long these will take. There are three options:

- Continue with the design as planned until the change is approved, recognizing that the longer it takes to be approved the more work there may be to undo in the design.
- Incorporate the changes into the design as if they are approved. For this you must get assurances from the project team in writing that the paperwork to proceed is a formality and will be forthcoming.
- Put the design in question on hold until approval is received.

The golden rule is that without an approved trend or scope change, no changes are to be made. This is a very reasonable, necessary, and important requirement for the project, and all companies should have a procedure for submission and approval/rejection. Your company should have forms for trending and scope changes. They will not always have the same title at each company, but will be along the lines of "Engineering Notice of Trend" and "Scope Change Request."

1.4.11 As-Builting

It may seem unnecessary to consider as-builting at the beginning of the project, given that it is one of the last tasks to be completed. However, you will need to establish exactly what your client requires to have as-built in order to complete your manhour estimates. Your client may only require this of critical documents, and will define "critical" for you. Critical documents can be defined as documentation that government regulatory bodies and the company deem must be kept current for the continued safe operation of a facility.

While by no means an exhaustive list, the documents generally listed that involve the piping group are:

- P&IDs.
- Line Designation Tables (LDT).
- Plot plan and equipment location plans.
- Underground piping plans.
- Heat tracing (hot oil, steam, glycol).

It is possible that you may also be required to as-built the 3-D models and the piping arrangements.

The client procedure must list and give direction in the documents to be as-built.

Depending on whether your company or the client is handling the construction management, the client or company procedures must give direction in:

- How the as-built changes will be captured, e.g., redline markups in the field.
- How the as-builts are to be turned over to the engineering group.
- How the engineering company is to turn the completed as-built documentation over to the client.

Internally, at your engineering company, you must plan for:

- Reviewing the as-builts, i.e., that the noted changes that occurred during construction were documented and approved.
- Drafting and checking of the field collected as-built information.
- Sign-off.

1.4.12 Project Close-out

At the end of the project, a close-out will be required. This generally involves handing over all the models, databases and drawings to your engineering company's document control group for close-out with the client's document control group. Make sure that you have an internal close-out procedure and a close-out procedure from the client, and that you fully understand the requirements of both. If one or both of these has not been provided, you may need to develop them in conjunction with your own people and the client.

1.5 Piping Execution Plan

There are two key documents that are the basis for the project: the Design Basis Memorandum (DBM) and the Project Execution Plan (PEP). A piping execution plan and a design and drafting execution plan will form part of the PEP, and you as the piping lead are expected to be a leading contributor in the writing of these sections. The piping execution plan may be a section unto itself or it may be a subset of the design and drafting execution plan.

The piping execution plan and the drafting execution plan are where all the decisions that have been made are captured in writing. General decisions on drafting that involve all disciplines, such as CAD software, are captured in the design and drafting execution plan, and specific piping related decisions are captured in the piping execution plan.

The following is a brief discussion of the DBM and the PEP, as they are the guiding documentation for you and your designers and the basis for everything that is to follow.

1.5.1 Design Basis Memorandum (DBM)

The DBM defines the project scope and describes the technical basis for detailed engineering. The typical content of a DBM includes the following sections:

- Project Overview and Facility Description.
- Facility Design Basis, e.g., Specifications and Standards.
- Safety Design Basis.
- Process Design Basis.
- Civil and Structural Design Basis.
- Mechanical Design Basis.
- Electrical Design Basis.
- Instrumentation and Controls Design Basis.
- Any additional relevant basis information.

1.5.2 Project Execution Plan (PEP)

The PEP describes how the project will be executed and typically contains the following sections:

- Cost Estimate.
- Schedule.
- Organization Plan.
- Project Controls Plan.
- Quality Assurance Plan.
- Safety and Health Plan.
- Regulatory Compliance Plan.
- IT Plan.
- Contracting and Procurement Plan.
- Document Control Plan.
- Engineering Execution Plan.
- Design and Drafting Execution Plan.
- Construction Execution Plan.
- Commissioning Plan.

Your designers must read and understand these documents because they constitute the official instruction for the project that everyone has to follow. Two key sections in the PEP are the Contracting and Procurement Plan and the Construction Execution Plan.

1.5.3 Contracting and Procurement Plan and Construction Execution Plan

The contracting and procurement plan and the construction execution plan outline the procurement and construction philosophies and are the basis from which many other decisions and planning by the piping lead will stem:

- Modularized and field erected piping splits.
- Construction Work Package (CWP) boundaries.
- Model boundaries.
- Shop and field material splits.
- Procurement splits.
- CWP drawing packages.
- Scopes of Work (SOW).

These documents may be in a preliminary stage when you first start and you may have to help define the contents. Failure to clearly establish the procurement and construction execution plans at an early stage and starting into detailed design without them can result in material, modeling, and drawing boundaries that may not later match the final desired CWP breakdown. This can lead to confusion for the fabricators and erectors, or added hours to rectify the splits.

1.6 Conclusion

You must investigate and secure all of the above, and should any of the above not be readily available, you must raise the flags with your project management team as to where it may be secured from and adopted for the project.

There is a lot to consider, and many questions to be answered, and much will be discussed in more detail in the subsequent chapters.

CHAPTER 2

Procedures

2.1 Introduction

In this chapter we will be discussing the key procedures that are required. You must have the means to monitor the progress of your project and capture the changes. You cannot possibly manage the work if you do not have a clear, concise plan to do so. It will become apparent very quickly that many of these procedures involve a paper trail. Even in today's computerized environment, paper is still a wonderful medium for communication and record keeping.

The following procedures are examples explained in detail. You may choose to follow them as presented, make your own modifications to suit your needs, or develop your own procedures. The important thing to remember is that a procedure is required. The tracking of the progress and the communications between the disciplines are made more efficient when everyone follows a procedure.

2.2 Master Stick Files, Working Copies, and Inter-discipline Drawing Reviews (IDR)

Stick files, working copies, and inter-discipline drawing reviews (IDRs) are closely interrelated. The use of stick files is backed up with working copies, and the IDRs are a periodic reality check of where the project is in production.

These procedures provide a needed "paper trail" that has somewhat fallen by the wayside with the advent of CAD. During the manual drafting days when drawings were filed in drawers, designers did not pull the drawings out every day to update the latest comments from an engineer. Collecting multiple mark-ups on multiple copies of the same drawing was a recipe for things to get lost. Consequently, a master stick file was used as a depository for all comments to be captured on until the right time to update the drawings was decided.

With CAD and electronic files on a network, anyone can run as many plots as they want, and designers can jump into and out of many drawings in a given day. A tendency has developed towards printing copies, marking-up and giving them to the designer for drafting, sometimes on a daily basis. Yet, the same questions that needed to be answered and controlled years ago are still in need of the same today. What was changed? Who made the changes? When did they make the changes? Why did they make the changes? And, who approved the changes?

The use of 3-D CAD over 2-D CAD doesn't help with these questions. Some would argue that everything is in the model and approved during model reviews. While model reviews are an important aspect of 3-D design and the printed model review comments/screen shots certainly leave a paper trail, the reality is that all questions are not answered, nor are they apparent when viewing the models. Having a design change captured in a model does not explain why the change was made. The end product is still a drawing, and managing the drawings is still the best way of controlling your project, for many reasons that will be explained below.

There was a time when all drawings followed the master stick file/IDR procedure from the beginning to the end of the project. Now this primarily applies to the P&IDs, plot plans, equipment location plans and key plans. With the advent of 3-D CAD, piping arrangements and isometrics are now generated towards the end of the project, and they will no longer follow this path until they are ready for checking. 3-D models have become the communication tool between the disciplines for change during the design stage. For instance, the piping group often places "temporary steel" in the 3-D models for the structural designers to follow and no longer mark up a stick file. However, the piping arrangements and isometrics must be circulated as an IDR at the checking stage and it is during this time that the stick files can be created. Beyond IFC the stick files are required during construction for reference and to capture the resolutions/changes made when answering questions from the fabrication and construction contractors.

2.2.1 Master Stick Files

The benefits of master stick files are:

- Provides one location for the capturing and recording of changes (what/who/when/why), unlike verbal instruction, e-mails, marked-up bootleg copies of drawings, and post-it notes which are difficult, if not impossible to track.
- All team members know where to go to get the latest information. The latest copy of a drawing being the one on the stick-file and not, if people are following the procedure, a mark-up sitting in someone's office or the electronic file on the network.
- Consolidation of effort. All team members can see what the others are planning, how they are being impacted and how they may be impacting others.
- Provides for a managed drawing update. Instead of engineers showing up with mark-ups at indiscriminate times and interrupting the workflow, comments are captured on an ongoing basis until the project engineer and the piping lead decide an update is in order.
- Provides a historical and permanent record of changes throughout the life of the project.

2.2.1.1 Storing Master Stick Files

Stick files are an assembled collection of the project drawings. Each drafting discipline must maintain their drawings in their area and divide them according to the drawing type. For the Piping Group these are primarily the:

- P&IDs.
- Line Designation Tables (LDT).
- Plot plans, equipment location plans, and key plans.
- Piping arrangements.
- Isometrics.

A note about the P&IDs: depending on your company's size and/or the structure of the engineering departments, you may not have custody of the P&IDs until later in the project. In many companies the process group has their own drafting support until after IFE. If the responsibility for the drafting comes under the piping lead, or once the custody has been transferred to the piping group, then you should manage them in a stick file. Hopefully your process group, and the other departments for that matter, will do so also.

The master stick files can be stored in a number of different ways:

- Full size drawings held in the traditional stick file clamp and hanging arrangement.
- Full or reduced size drawings taped to a roller board.
- Reduced size drawings maintained in oversize three ring binders.

I like to place P&IDs and plot plans, etc., in hanging stick files or on a roller board, and piping arrangements in hanging stick files and isometrics in the oversize three ring binders. I also prefer the full size drawings, as they have lots of room for mark-ups and are to scale in the case of scaled drawings.

2.2.1.2 Managing Master Stick Files

These project masters are always in the custody of the piping lead and managed in the following ways (refer also to Figure 2–1 Stick File Flowchart):

- Each stick file has a cover sheet and a drawing index.
- All the drawings will be stamped 'MASTER STICK FILE' and dated.
- Drawings are organized in numerical order.
- The stick files are not to be removed from the drafting area without the piping lead's approval.

  Be very guarded against letting your stick files leave the area. One, because they may not come back, and two, because while they are gone someone else may come to do a mark-up and you would have to send them on a search mission, which is a good way to lose their goodwill to follow your procedure.

- The piping lead must be informed of all changes.
- A stick file log is created (spreadsheet) to be filled in by everyone who marks-up changes on the stick file. The log can reside on the front of the stick file or in a separate binder close by. The log requires the following columns:
  - Printed Name.
  - Signature.

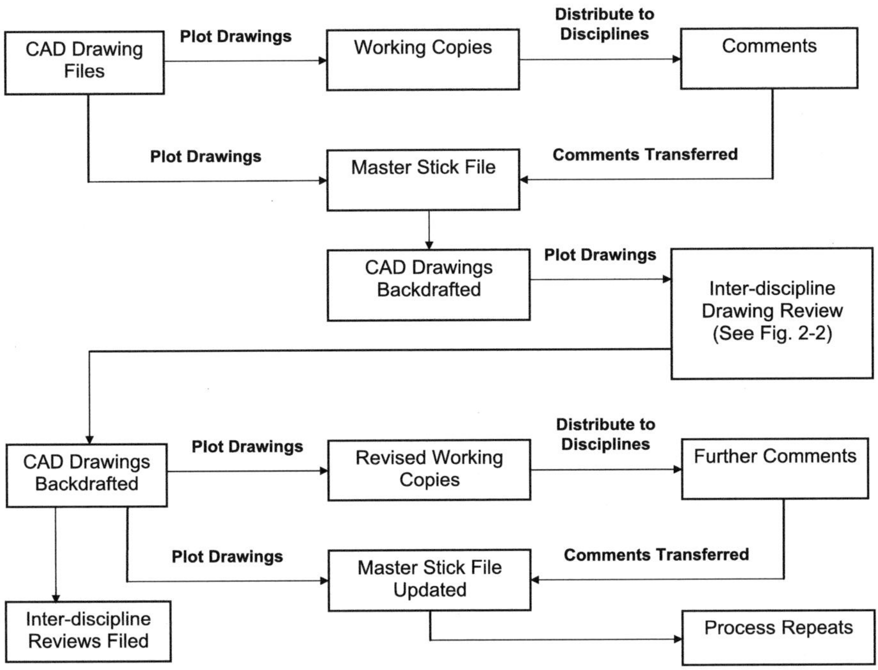

Figure 2–1 *Stick File Flowchart.*

- Date.
- Drawing Number.
- Revision Number.
- Reason For Mark-Up.

- All mark-ups must be legible. Changes are to be signed and dated.
- All project team members must mark-up the stick file with any changes/additions, including holds, which can affect the ongoing engineering work. Preferably you will have them do their mark-ups in a discipline color code. The color coding helps to establish who did the mark-up when the author forgets to sign or the signature is illegible. See Table 2–1 Discipline Mark-Up Color Codes.
- Questions on stick files are not allowed. Questions must be discussed with the appropriate party and the resolution marked-up. This is so that when the time comes to do the backdrafting, the drafter is not spending piping hours chasing others to get the answers.

Table 2–1 Discipline Mark-up Color Codes

	STICK FILES AND INTERDISCIPLINES (Note 1)	CHECK PRINTS (Note 3)
RED	MARK-UPS BY PIPING DESIGN	CHECKER MARKING REQUIRED CHANGES AND ADDITIONS
YELLOW	BACKDRAFTER HIGHLIGHTS BACKDRAFTED COMMENTS	CHECKER INDICATING CORRECT INFORMATION
FLESH		CHECKER MARKING ITEMS TO BE DELETED (Note 2)
BLACK	MARK-UPS BY CIVIL/STRUC-TURAL DESIGN AND ENGINEERING	CHECKER'S NOTES TO BACKDRAFTER
BLUE	MARK-UPS BY ELECTRICAL DESIGN AND ENGINEERING	CHECK MARK BY BACK-DRAFTER INDICATING CHECKER'S MARK-UPS INCORPORATED ON CAD FILE
GREEN	MARK-UPS BY INSTRUMEN-TATION AND CONTROLS ENGINEERING	CHECK MARK BY CHECKER INDICATING BACKDRAFTER INCORPORATED THE INFOR-MATION CORRECTLY
BROWN	MARK-UPS BY MECHANICAL AND STRESS ENGINEERING	
PURPLE	MARK-UPS BY PROCESS ENGINEERING	
ORANGE	MARK-UPS BY PROJECT ENGINEERING	

Notes:
1. Sign and date mark-ups.
2. Do not obliterate the existing information to be removed.
3. Checker and backdrafter to sign and date checkprints.

- The piping lead will have the comments on the stick file drawings backdrafted, either prior to a scheduled IDR or by monitoring the mark-ups and periodically calling for an update. Changes that have an impact on cost and/or schedule will be discussed with the project engineer prior to backdrafting. Trends and scope changes must be approved prior to the work being carried out.
- Further mark-ups are not allowed on the stick file during the backdrafting exercise as this just interrupts the backdrafting effort. Request that new mark-ups be held until the IDR set is issued.
- When the stick file is updated the previous revisions will be stamped 'SUPERSEDED.'
- Superseded stick file copies will be filed and retained for the duration of the project.

2.2.2 Working Copies

Working copies:

- Provide a personal copy to record on.
- Reflect the status and decisions to date from last inter-discipline drawing review.

2.2.2.1 Managing Working Copies

Working copies are issued to the project team members to ensure that everyone has copies of the latest drawings to refer to:

- Working copies will be ANSI B (11" × 17") or ISO A3 (297mm × 420mm), stamped 'WORKING COPY' and dated.
- The piping lead will issue the first set of working copies to the team members as soon as the drawings are ready for general distribution. The project engineer and the piping lead will agree which team members are to receive working copies. As a minimum this would be the same individuals who are to be included on the IDR. Subsequent issues of working copies by the piping lead will be after a stick file drawing update or after an IDR.
- Team members must periodically refer to the stick file and update their working copies by recording all comments.

A cautionary note: comments involving a trend or a scope change can only be acted upon once approved.

2.2.3 Inter-discipline Drawing Reviews

The benefits of IDRs are:

- Ensures that all team members have signed-off on the design to date.
- Ensures trends and scope changes are addressed in a timely and orderly manner.

Inter-discipline reviews are conducted to ensure that all disciplines are aware of the developments and have opportunity to comment on those aspects that affect their work. Also, parties that have direct input into the design are satisfied that their communications have been understood and incorporated correctly, and that the drawings reflect the latest available information. The IDR ensures that all team members sign-off periodically during design development.

IDRs may be conducted at any time at the discretion of the project engineer, but as a minimum reviews will be conducted prior to a major issue, i.e.,

- Issued For Review (IFR).
- Issued For Approval (IFA).
- Issued For Engineering (IFE) also known as Issued For Design (IFD).
- Issued For Construction (IFC).

By agreement between the project engineer and lead discipline engineer, the IDR will be conducted in one of two ways:

- A review meeting with all parties present.

or

- Mark-up of a review set of drawings located in a central area by a set deadline.

Commonly this will depend on the urgency to complete the review.

Note that some companies choose to circulate the IDR from one reviewer to the next rather than place them in a central review area; however, the net result is the same.

2.2.3.1 Managing Inter-discipline Drawing Reviews

The project engineer is ultimately responsible for ensuring the timely review of the documents and drawings and the completion of reviews, but the success of the reviews depend upon a joint effort of the project engineer, discipline engineer, and piping lead, each following the below prescribed point form steps in a chronological order.

Referring to the attached flowchart, Figures 2–2a and 2–2b, the drawings are prepared for the review.

- The project engineer, discipline engineer and piping lead:
  - Agree to the issue date and the drawings that are to be included in the IDR.
  - Inform the team members that an IDR is imminent. Team members are given an appropriate amount of time to get any last minute comments marked-up on the master stick file.
- The piping lead:
  - Has the drawings revised and checked to the mark-ups on the master stick file.
  - Assembles one full size set for review.
  - Stamps the full size set with the inter-discipline drawing review stamp, Figure 2–3, and forwards the set to the project engineer.
- The project engineer:
  - Decides who is required to review the drawings. (Table 2–2 is a suggested matrix of the types of drawings to be reviewed and who should review them. The final decision on the drawings to be reviewed and those required to review them rests with the project manager, project engineers, and the lead discipline engineers.)
  - Fills in the inter-discipline review stamp.
  - Crosses off the departments that are not required to review the drawings.
  - If deemed necessary, prioritizes the order of the reviewers in the CIRC. (circulation order) column, i.e., 1st, 2nd, 3rd, etc. Note: The project engineer, the "Issuer," will always be the first and the last reviewer. The discipline engineer should always be the second to last reviewer.
  - Notes the date that the review must be completed by, and who the mark-ups are to be returned to.

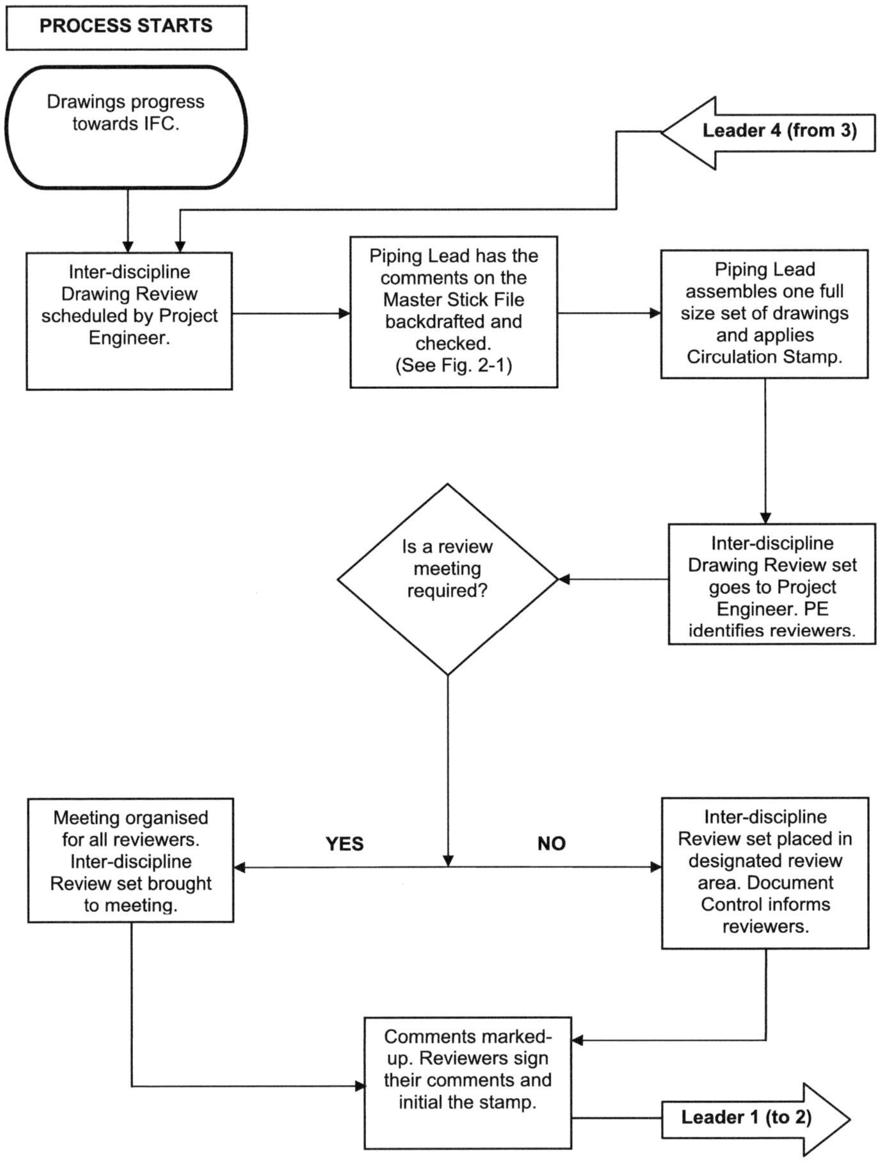

Figure 2–2a *Inter-discipline Drawing Review Flowchart.*

- Informs document control to place the full size set of drawings in the designated review area and to notify the reviewers or organizes a review meeting. Note that if there is to be a meeting, it is a good idea to bring along the master stick file for reference.

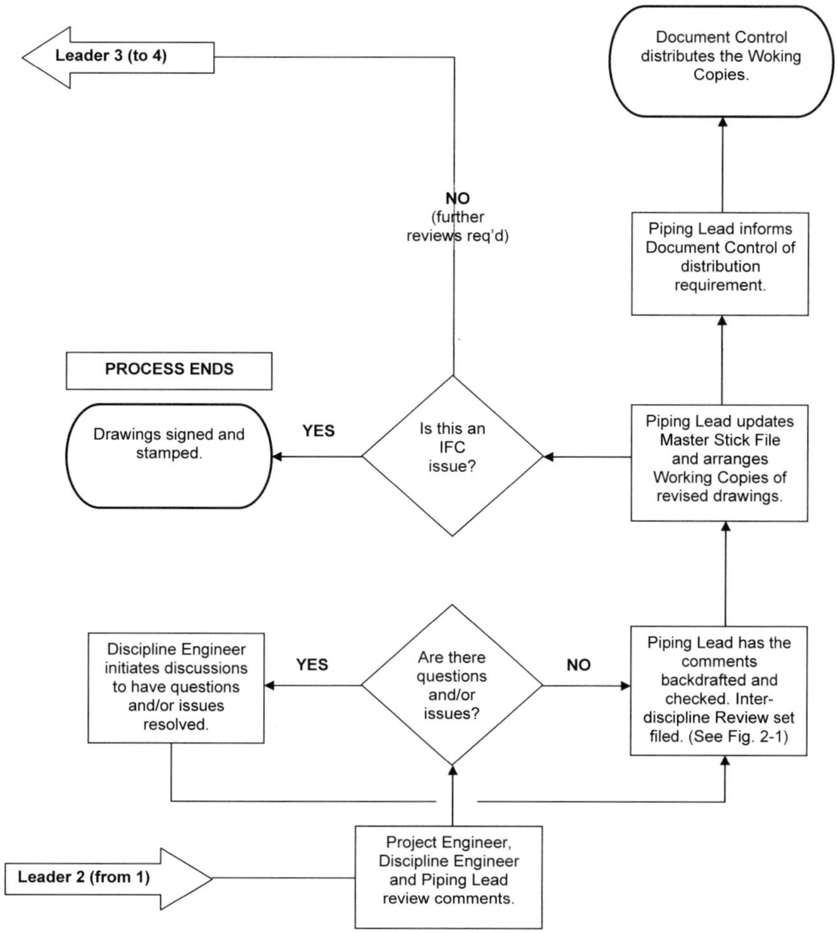

Figure 2–2b *Inter-discipline Drawing Review Flowchart. (continued)*

- The project engineer, discipline engineer and piping lead (after the reviewers; see final bullet):
  - Review the comments.
  - Resolve any unresolved issues or questions. It is incumbent upon these individuals to resolve any issues, as it is not appropriate to have a designer tackle unclear or conflicting instruction during backdrafting.
  - Discuss comments affecting cost and/or schedule. The project engineer either approves or rejects the comments. Rejected mark-ups are communicated back to the originating reviewer.

COMPANY NAME Inter-discipline Drawing Review			
CIRC.	DEPARTMENT	INITIAL	DATE
	Issuer		
	Process		
	Mechanical		
	Piping Stress		
	Piping		
	Civil		
	Structural		
	Electrical		
	Instrumentation		
	Controls		
	Construction		
	Operations Sponsor		
	Project Engineer		
Mark-ups to be in Discipline Colour Code			
Reviews to be completed by:		Return to:	

Figure 2–3 *Inter-discipline Drawing Review Stamp.*

- The piping lead:
    - Has the comments backdrafted immediately.
    - Updates the master stick file.
    - Issues each team member with an updated working copy.
    - Files the IDR set for the duration of the project. At the project closeout these will be archived with other project documentation.
- The reviewers:
    - Review that their comments placed on the master stick file have been correctly incorporated into the IDR set of drawings.
    - Add new comments.
    - Sign and date the interdisciplinary review stamp on each drawing.

Table 2–2 Inter-discipline Drawing Review Matrix

LEGEND: R = Responsible for final drawing S = Must review and sign-off the drawing I = Issued for information only	PROJECT MANAGER	PROJECT ENGINEER	PROCESS	MECHANICAL / PIPING	PIPING STRESS	CIVIL / STRUCTURAL	ELECTRICAL	INSTRUMENTATION	CONTROLS	CONSTRUCTION	OPERATIONS SPONSOR		
PROCESS:													
Process Flow Diagrams (PFD)	S	S	R	I			I	S	S		S		
Piping and Instrumentation Diagrams (P&ID)	S	S	R	S	I		S	S	S		S		
Line Lists	S	S	R	S	I		I	I	I		I		
MECHANICAL / PIPING:													
Plot Plan	S	S	S	R	I	S	S	S	S	S	S		
Piping Plans, Sections and Details		S	S	R	S	S	S	S	S	S	S		
Piping Isometrics				R	S					S			
Underground Piping Plans		S	S	R	S	S				S	S		
CIVIL:													
Grading Plans and Details	I	S		S		R				S			
STRUCTURAL:													
Structural Plans, Sections and Details		S		S		R	S			S			
Foundation Details		S		S		R				S			
Piling Plans				S		R				I			
Building Floor Plans and Elevations		S		S		R	S			I	S		
ELECTRICAL / I&C:													
Area Classification and Grounding Layout	S	S		S			R	S	S		S		
Cable Tray Layouts				S		S	R			S			
Electrical and Instrumentation Location Layouts				S			R	S	S		S		
Single Line Diagrams		S					R						
Alarm and Shutdown Keys	S	S	S				S	I	R		S		

- Mark-up per the discipline color code.
- Mark-ups must be legible and signed. If the signature is not discernible, then the reviewer should also print his/her name using block capitals.
- All comments must be marked-up on the IDR set. It is not allowed to staple the working copy to this set.
- Discuss questions with the appropriate reviewer(s) and reach a resolution. Involve the project engineer, design engineer, and piping lead if necessary.

- Review the newly issued working copies to ensure the comments placed on the IDR set have been correctly interpreted. Inform the piping lead of any errors and mark these up again on the newly updated master stick file.

2.3 Vendor Drawing Reviews

Vendor document control is a complex procedure starting with the engineering company data sheets, followed by the requisitions, purchase orders, and ending with the certified as-builts and vendor data books. As it is not the purpose of this book to get into the details of the procedures of other departments, except in context with the responsibilities of the piping lead, we will limit our discussion on this subject.

The vendor drawing review procedure is just like the IDR procedure and requires the same input from the project and discipline engineers in conjunction with document control. Unlike the IDR where the piping lead is key to the creation, integrity, and moving the piping discipline drawings through the system, the lead's role for the vendor drawings is one of managing them once they come into the piping group. Figure 2–4 is a simplified flowchart of the vendor document review procedure.

As per the piping drawings, a master stick file is required for all disciplines to mark-up their comments on during the period of time a drawing has been returned to the vendor and the vendor resubmits the drawing with the required changes. There are systems in place in the piping group to govern the integrity of all documentation handled. Specifically the piping lead will manage the mechanical vendor drawings in the following ways:

- Create a storage place for the master stick file copies in the piping area. Usually this is going to be in a filing cabinet with file folders, each of which will have a tab with the equipment number. Because vendor drawings come in a variety of sizes, even from the same vendor for a given piece of equipment, this is by far the best way.
- When vendor drawings are circulated for review, ensure that the applicable piping members complete their reviews in a timely manner.
- Review the mark-ups from all team members. Mark-ups may be new information, previous mark-ups not incorporated by

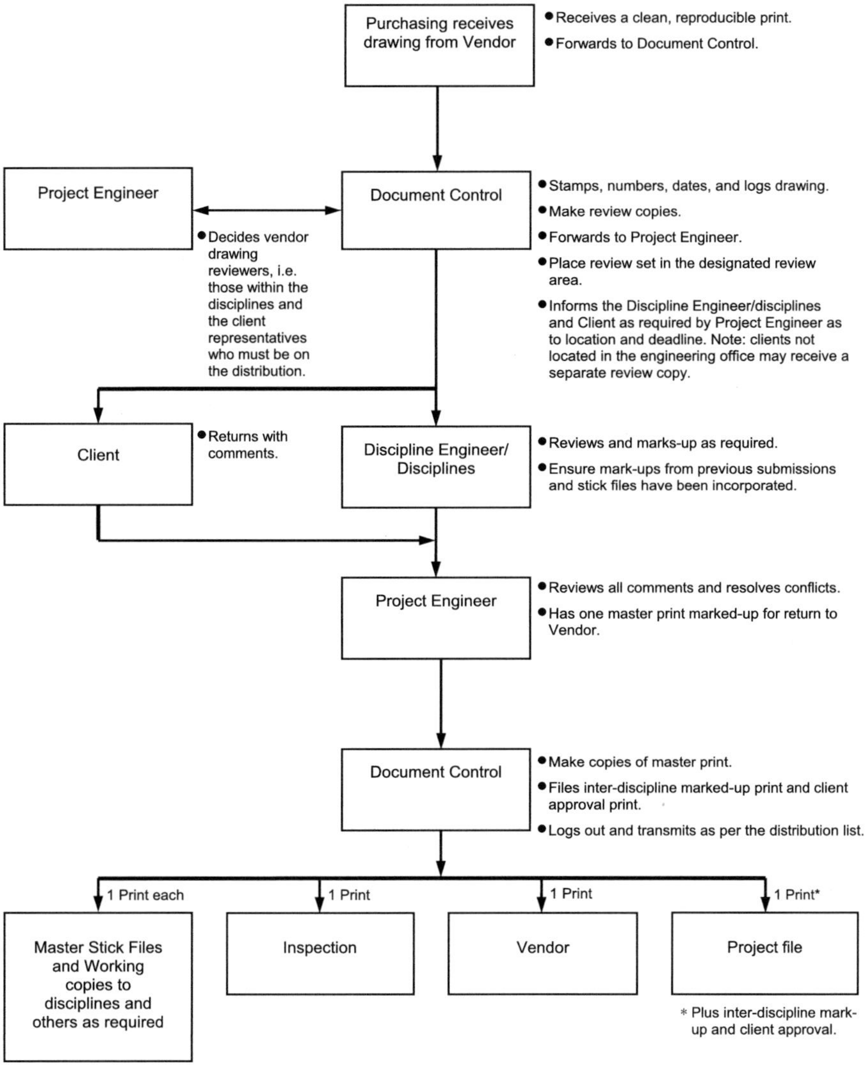

Figure 2–4 *Vendor Document Review Flowchart.*

the vendor, and mark-ups from the stick files not yet communicated to the vendor.

- Discuss issues with the project engineer.
- Act as agent for the project engineer to have one clean mark-up created from the vendor review copy for return to vendor.
- Return the clean mark-up and marked-up review copy to document control.

- Maintain the master stick file by filing the latest master stick file copy and superseding the previous stick file copy.
- Secure working copies for the applicable piping team members.
- Bring mark-ups on the master stick to the attention of the project engineer. For instance, if a nozzle is moved there is no point in waiting until the next vendor submission to capture this change. It must be discussed with the vendor as soon as possible.

2.4 Line Numbering

Once the P&IDs have been developed to a reasonable stage of completion, line numbers should be assigned. This needs to happen prior to the start of 3-D modelling. Some companies will leave this task to the process group. I believe that it is one best handled by the piping lead. Generally speaking, a piping lead will have a better understanding of the piping layout, isometric production, and construction requirements. This information, commonly understood by the piping lead, is valuable when assigning line numbers. Consistency of approach to line numbering is a requirement in order to:

- Minimize the quantity of line numbers.
- Ensure that the grouping of lines in the LDT is by commodity and system.
- Assist construction. Line numbers are used for isometric drawing numbers, and therefore there is a direct relationship between line numbering and the order of arrangement of isometrics in the Construction Work Packages (CWPs). By using a structured approach to line numbering, the isometrics will be arranged such that the associated lines to be welded together will be grouped together.
- Avoid unnecessary and unwanted breaks in the isometrics. The software that automatically generates isometrics will make a break at all line number changes unless manually overridden. This may result in extra isometrics and can lead to extra field welds. Spooling fabricators often interpret all breaks between isometrics to mean a field weld location.

2.4.1 Line Numbering Rules

The following rules will help achieve the needed consistency:

- Rule 1—Assign numbers in the direction of flow as closely as possible for each P&ID, until all lines, including branches, are numbered. Do not forget the numbering of all miscellaneous lines at each piece of equipment (drains, vents, etc.). Start with the main process feed line, following the primary process. Secondary processes should be completed next. Utility lines should be numbered next, completing one utility system at a time. Never re-use deleted line numbers.
- Rule 2—Use one number for a line that runs from one piece of equipment to another in the direction of flow. The number may include one line size or more. List the different NPS pipe sizes separately on the LDT. Refer to Figures 2–5 and 2–12.
- Rule 3—Assign a new number whenever the pipe specification changes along the length of a line. Refer to Figures 2–6 and Figure 2–8(a).
- Rule 4—If the line class changes downstream of a control valve to a lower line class, carry the higher line class and number through to include the bypass valve and the block valve after the control valve. Refer to Figure 2–8(b).
- Rule 5—Sub-headers and branch lines to or from main headers will change number. Refer to Figures 2–5 and 2–6.
- Rule 6—Vents, drains, sample connections and other very short connections are an integral part of the line and will carry the same number as that line. The size must be noted on the P&ID. Refer to Figure 2–7.
- Rule 7—Lines that are close coupled to relief valves shall be assigned the same numbers as the headers from which they branch. Lines running to relief valves that are located some distance away from the header shall be assigned a new line number. If in doubt assign a line number. It can easily be deleted later if not required. Lines running from a relief valve to a collection header or to atmosphere will be assigned a new number. Refer to Figures 2–9 and 2–11.
- Rule 8—Long by-passes around vessels or other equipment shall be assigned a new number. Short by-passes between lines, such as those at control valve stations and relief valves are not given a new number. Refer to Figures 2–7(b), 2–8(b), 2–9(b), 2–10(b), and 2–11(b).
- Rule 9—Refer to Figure 2–10 when assigning numbers to relief valves on equipment.

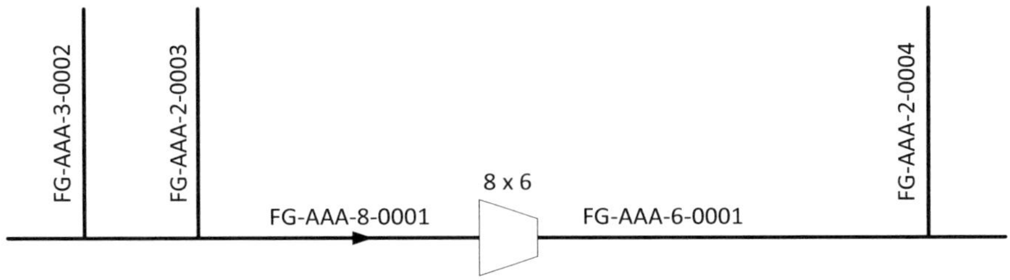

Figure 2–5 *Header with Line Reduction.*

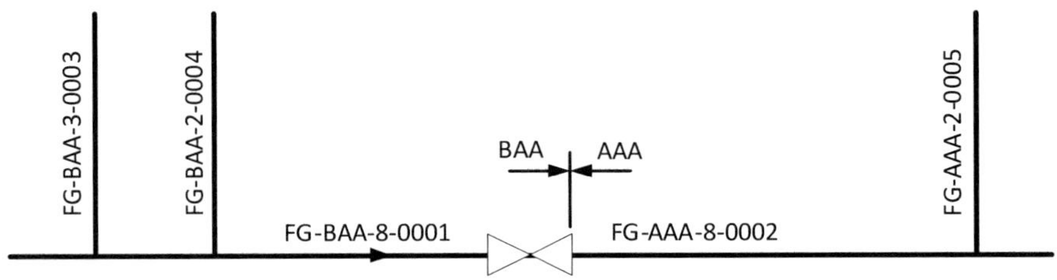

Figure 2–6 *Header with Spec. Break.*

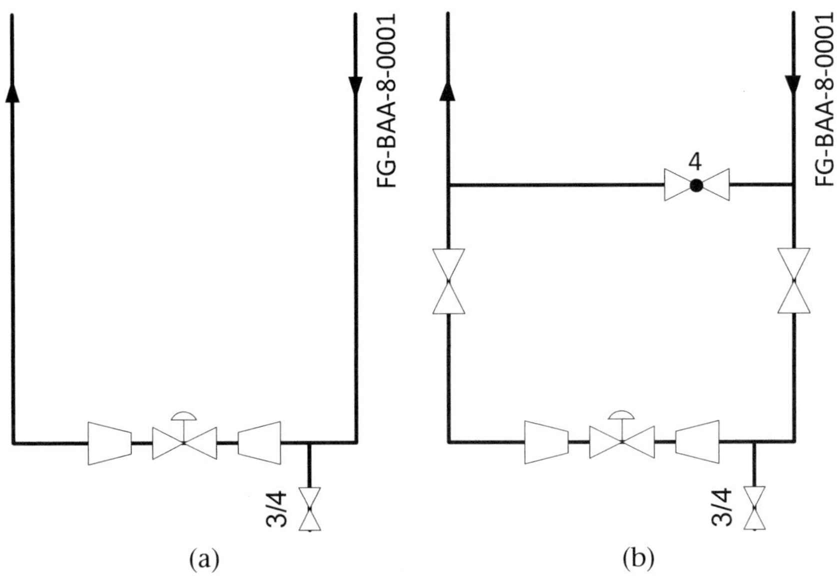

Figure 2–7 *(a) Control Valve without Bypass or Spec. Break. (b) Control Valve with Bypass without Spec. Break.*

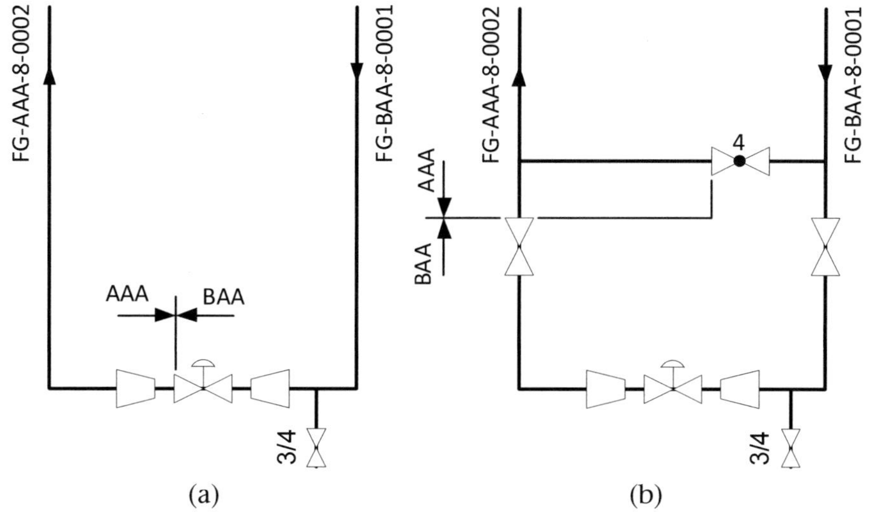

Figure 2–8 *(a) Control Valve without Bypass with Spec. Break. (b) Control Valve with Bypass and Spec. Break.*

- Rule 10—Where pumps, exchangers and similar equipment are piped in parallel, the incoming line number shall be continued through to the 'A' designated equipment. The outgoing line shall be assigned a new number. Refer to Figure 2–13.
- Rule 11—Header and manifold piping for aerial coolers shall be numbered as shown in Figure 2–14.
- Rule 12—In multi-unit plants, the line numbers often include a unit identifier, e.g., Unit 2, line number 2001. In these cases the line number should be maintained from the start to the termination, regardless of the number of units through which it passes. Branch lines are numbered according to the unit that they serve.

2.5 Stress Analysis

It goes without saying that stress analysis is an integral part of the overall piping design effort, but what is stress analysis? It is a term applied to the calculations that address piping loads resulting from: gravity, temperature, pressure, fluid flow, seismic activity, wind, and other environmental conditions. Codes establish the minimum scope of stress analysis to ensure stress limits are not exceeded. But that is not the sum of it. Because this is an extremely important component

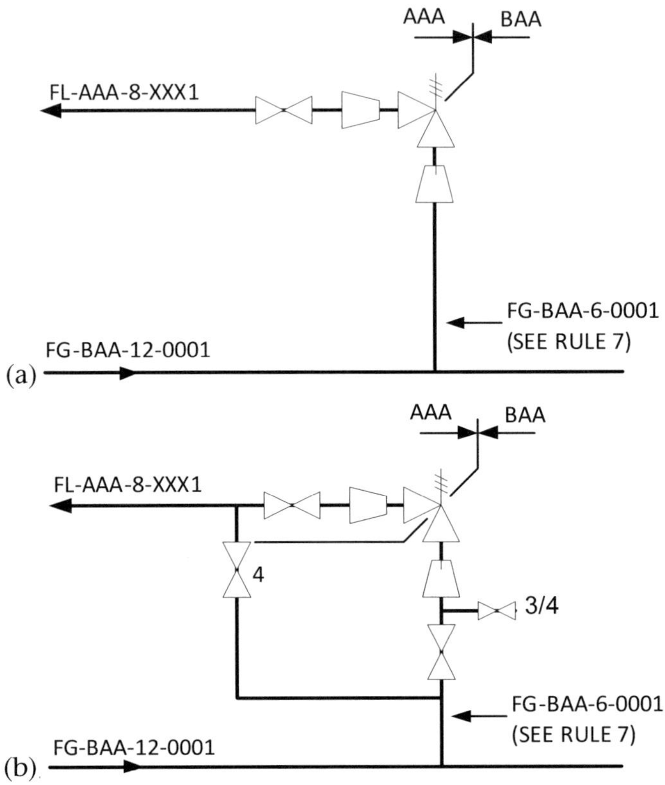

Figure 2–9 *(a) Single PSV without Bypass. (b) Single PSV with Bypass.*

of piping design there are requirements that this activity must be documented and the paperwork be archived. We need to manage the effort appropriately to ensure that the stress analysis is carried out in an efficient manner.

Piping stress analysis is highly interrelated with piping layout and support design. Establishing stress analysis requirements and monitoring the progress throughout the project avoids poor timing and reworking, and minimizes the possibility that something is missed.

2.5.1 The Stress Analysis Procedure

Your procedure must provide checks and balances. It is necessary that the stress analysis for all lines be considered at the beginning and throughout the life of a project, that the stress engineer's requirements are incorporated into the layouts, and that the history of this activity can be recreated. This requires the stress group to be involved

(a)

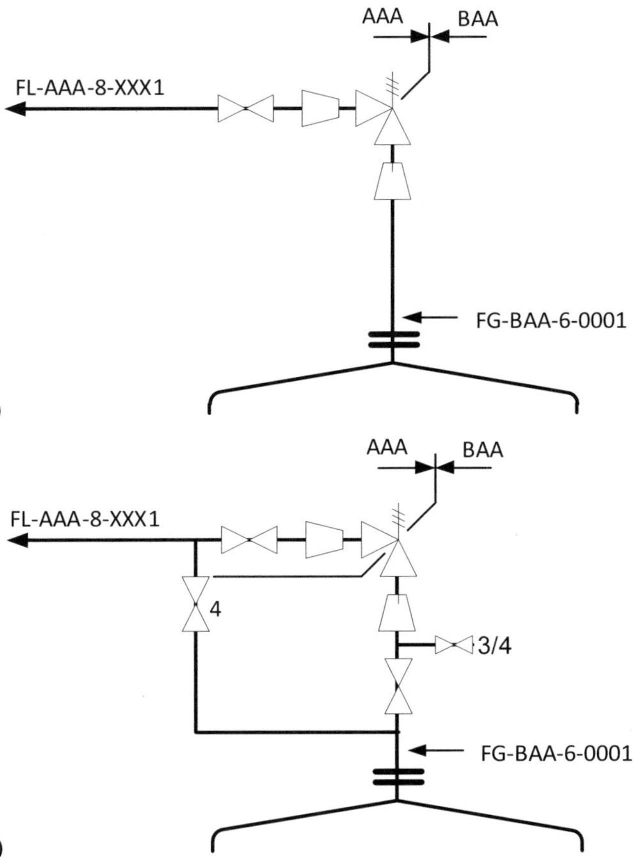

(b)

Figure 2–10 *(a) Single PSV without Bypass on Vessel. (b) Single PSV with Bypass on Vessel.*

in the piping layouts at the earliest possible opportunity, and that the flow of information between the two groups will be monitored.

What are these requirements and why?

- Engage the stress group prior to starting the piping layouts to establish the stress analysis criteria, i.e., methods of calculation: visual, manual calculations or computer analysis. Preferably you will have IFE P&IDs and LDTs to work with, but you may have to begin this exercise with IFA level documents. This establishes a level of priority. The lines indicated as requiring a computer analysis (Critical Lines) are the high priority lines, those that will require the most attention and the routings of which are most likely to be problematic. These are

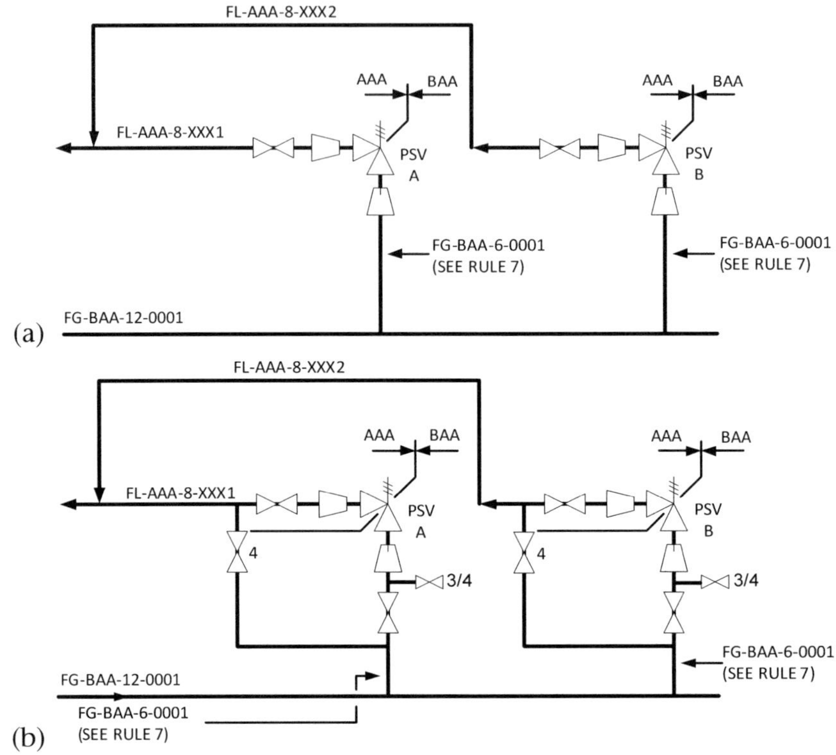

Figure 2–11 *(a) Multiple PSV without Bypass. (b) Multiple PSV with Bypass.*

the lines to be laid out first and submitted for stress analysis, commonly during the study stage of the design. Early review of these lines will save rework and/or costly changes later. Refer to Table 2–3 Stress Analysis Initial Determination Matrix.

- Identify the method of analysis in the LDT. This gives one location for reference for all interested parties. Refer to Figure 2–15 Stress Analysis Method listed in LDT.
- Establish the method for the flow of information from the piping group to the stress group and back. This will be by hard copy stress isometrics and may also be accompanied by the required file format for importation into the stress analysis software. Stress isometrics are submitted for each of the computer analyzed lines because formal stress analysis is done line by line. The hard copies of the isometrics provide the best record of the progress and convenience for mark-ups. These mark-ups

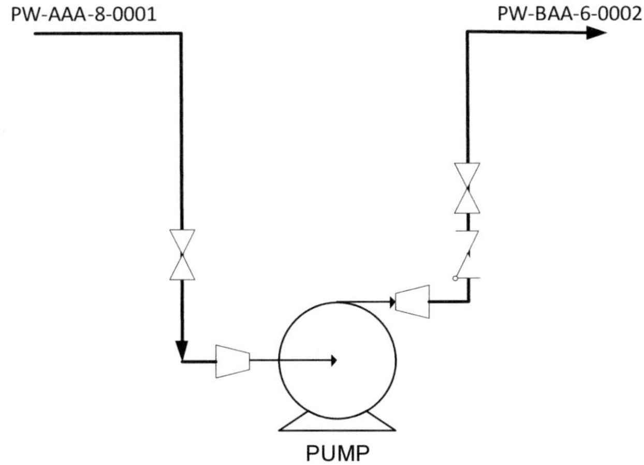

Figure 2–12 *Inlet or Suction and Outlet or Discharge for Single Piece of Equipment.*

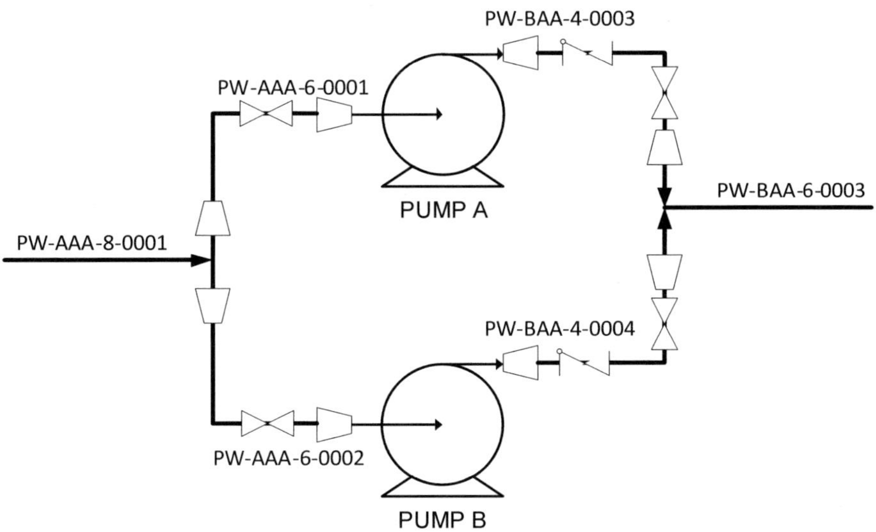

Figure 2–13 *Inlet or Suction and Outlet or Discharge for Multiple Equipment in Parallel.*

will become a document of record. Lines may go through several revisions before approval. It is important to know the status of all lines at any given moment, i.e., lines are in a state of flux; while some are approved, others are awaiting submission or resubmission to stress, and others are still being analyzed by

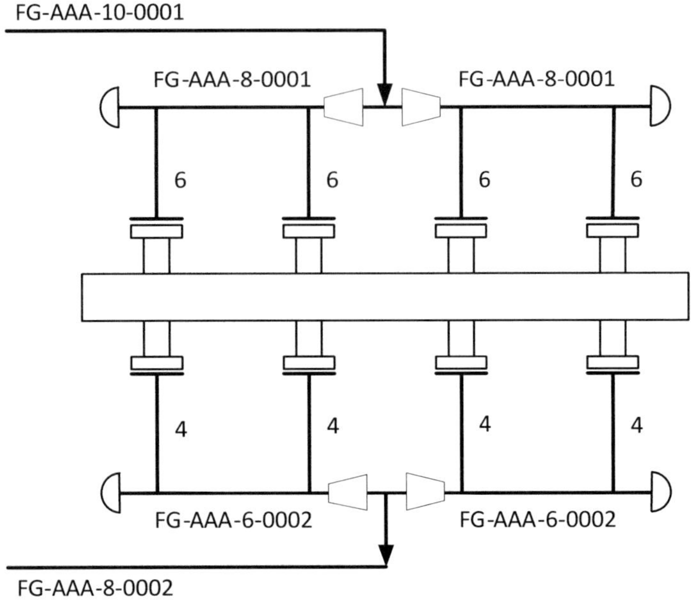

Figure 2–14 *Inlet and Outlet for Aerial Coolers.*

stress and are awaiting return to piping. The piping checker and signing engineer must have a means to satisfy themselves that they have checked the piping arrangements and the isometrics against the final approved information. It may also be necessary to recreate the stress analysis history of a given line or lines at a later date. Refer to Figure 2–17 Stress Analysis Flowchart.

- Monitor the flow of information. Set-up a stress log to track the submissions and resubmissions of stress isometrics to stress, and stress isometrics returned from stress. Refer to Figure 2–16 Stress Log.
- Store the stress isometrics. Each plant area needs two stress binders titled: "SUBMITTED TO STRESS" and "RETURNED FROM STRESS." These are to be maintained in a common area where they are accessible to the designers, checkers, and engineers. As with the stick files, they must not leave the area.
- Due to the importance of this information, and because on larger projects it will become a full time job, I strongly suggest

LINE DESIGNATION		REFERENCE	OPERATING CONDITIONS			DESIGN CONDITIONS						
						CODE		TEST		UPSET		EXPAN
ORIGIN	TERMINATION	P&ID DRAWING NUMBER	PRESS. kPa (GAUGE)	TEMP °C	STRESS ANALYSIS	PRESS kPa(g)	TEMP °C	PRESS kPa(g)	MEDIUM	PRESS kPa(g)	TEMP °C	TEMP °C

Indicates the method of stress analysis employed:
V = Visual
M = Manual
C = Computer ("Critical Lines", the minimum lines requiring a formal stress analysis. All other lines are to be reviewed during detailed design to determine if a formal stress analysis is necessary.)

Figure 2–15 *Stress Analysis Method listed in LDT.*

that there be a designated stress coordinator working on behalf of the piping lead. This individual is to be a single point of contact for moving stress isometrics to/from the stress group, keeping the stress log and binders up to date, and arranging working copies for the designers and checkers. If you leave it to each individual to do this, you will find very quickly that the tasks are not completed and that information will go missing.

Refer to Figure 2–17 Stress Analysis Flowchart and the 2.5.2 notes for the Stress Analysis Procedure.

STRESS ISOMETRIC NUMBER	LINE NUMBER(S) ON STRESS ISOMETRIC	REFERENCES		DRAWN BY		SENT TO STRESS	RETURN FROM STRESS	SENT TO STRESS	RETURN FROM STRESS	SENT TO STRESS	RETURN FROM STRESS	APPROVED	REMARKS
		P&ID No.	MODEL No.										
					REV								
					DATE								
					REV								
					DATE								
					REV								
					DATE								
					REV								
					DATE								
					REV								
					DATE								
					REV								
					DATE								
					REV								
					DATE								

Figure 2–16 *Stress Log.*

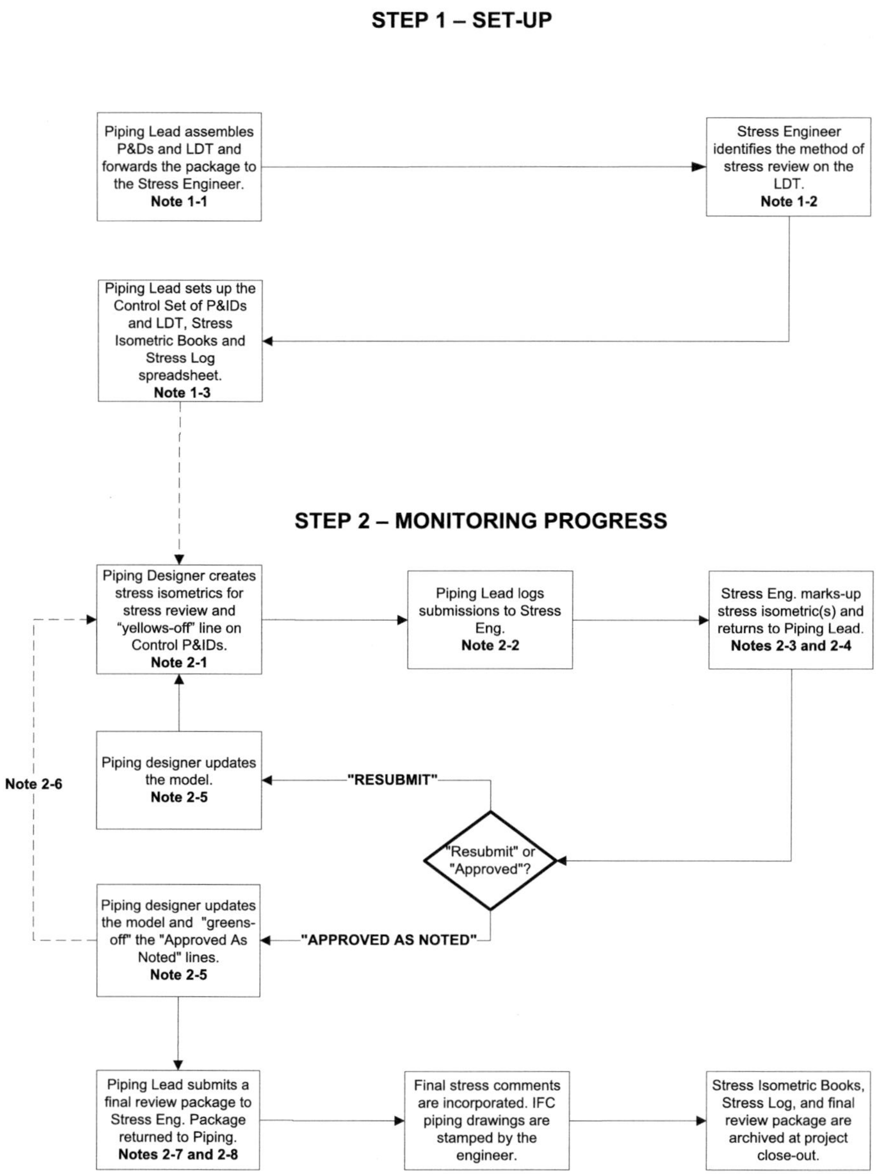

Figure 2–17 *Stress Analysis Flowchart.*

2.5.2 Stress Analysis Procedure Notes

2.5.2.1 Set-up

- Note 1-1—piping lead:
  - Assembles a Control Set of P&IDs, LDTs and forwards the package to the stress engineer.
- Note 1-2—stress engineer:
  - Identifies the stress review requirements in the LDT using Table 2–3 Stress Analysis Initial Determination Matrix, P&IDs, and his/her own judgement.
  - Returns the package to the piping lead.
- Note 1-3—piping lead:
  - Sets-up a "SUBMITTED TO STRESS" stress isometric book and a "RETURNED FROM STRESS" stress isometric book.
  - Stores the control set of the P&IDs and LDTs in the design area.
  - Places the stress log spreadsheet into the project directory stress folder.

2.5.2.2 Monitoring Progress

- Note 2-1—piping designer:
  - Identifies the lines for which he/she are responsible to create stress isometrics—by reference to the control set of P&IDs and LDTs.
  - Routes lines and submits stress isometrics on an ongoing basis.
  - Creates an electronic stress isometric drawing file and stress software importation file (assuming compatible software, otherwise this will not be possible) for each stress isometric and places them in the project directory stress folder. Use line number for isometric number and revisions starting at Rev. A. See Figure 2–18 Stress Isometric.
  - Prints two hard copies of each stress isometric, stamped and dated: one "PIPING COPY" and one "STRESS COPY." (Note that the stress copy stamp is to include two boxes for the stress engineer's use: "APPROVED AS NOTED" and "RESUBMIT.")
  - Issues both to the piping lead.

Table 2–3 Stress Analysis Inital Determination Matrix

NPS	Design Temperature (°C)											
	50	75	100	125	150	175	200	225	250	275	300	325
1	V	V	V	V	V	V	V	M	M	M	M	C
1½	V	V	V	V	V	V	V	M	M	M	C	C
2	V	V	V	V	V	V	M	M	M	C	C	C
3	V	V	V	V	V	M	M	M	C	C	C	C
4	V	V	V	V	M	M	M	C	C	C	C	C
6	V	V	V	M	M	M	C	C	C	C	C	C
8	V	V	M	M	M	M	C	C	C	C	C	C
10	V	M	M	M	M	C	C	C	C	C	C	C
12	M	M	M	M	M	C	C	C	C	C	C	C
14	M	M	M	M	C	C	C	C	C	C	C	C
16	M	M	M	M	C	C	C	C	C	C	C	C
18	M	M	M	C	C	C	C	C	C	C	C	C
20 +	M	M	M	C	C	C	C	C	C	C	C	C

NOTES:

Using the table above:

1. If a design temperature falls between two table design temperatures, use the higher of the two table design temperatures for determining the level of stress analysis.
2. A "V" in the table indicates that analysis may be possible by visual inspection of the piping arrangements. Indicate "V" on the Line Designation Table. A stress isometric is not required.
3. An "M" in the table indicates that analysis may be performed by manual or hand calculation methods. Indicate "M" on the Line Designation Table. A stress isometric will be required if requested by the stress group.
4. A "C" in the table indicates that analysis will be performed by computer program methods. Indicate "C" on the Line Designation Table. A stress isometric is required.

Other lines that may require analysis by computer program methods, even if not indicated as such by the above table, are:

1. All lines subject to vibration or pulsation.
2. All lines subject to severe cyclic service.
3. All lines connected to sensitive equipment such as:
   a. Centrifugal and reciprocating, compressors and pumps, with the exception of auxiliary piping.
   b. Steam lines to and from turbines.
   c. All air-cooled heat exchanger piping systems connecting to exchangers with four or more header box nozzles.
4. All lines to and from reactors.
5. All lines subject to steam purging (min. t = > 180 C)
6. All lines Cat. "M"
7. All lines subject to dynamic forces, e.g. pressure relief systems.
8. All jacketed piping.
9. Branch line tie-ins of large branch piping.
10. All underground lines where pressure > 1000 psi or > 70 C
11. All thin-walled piping NPS 18 or larger, having an outside diameter divided by wall thickness ratio of more than 90.

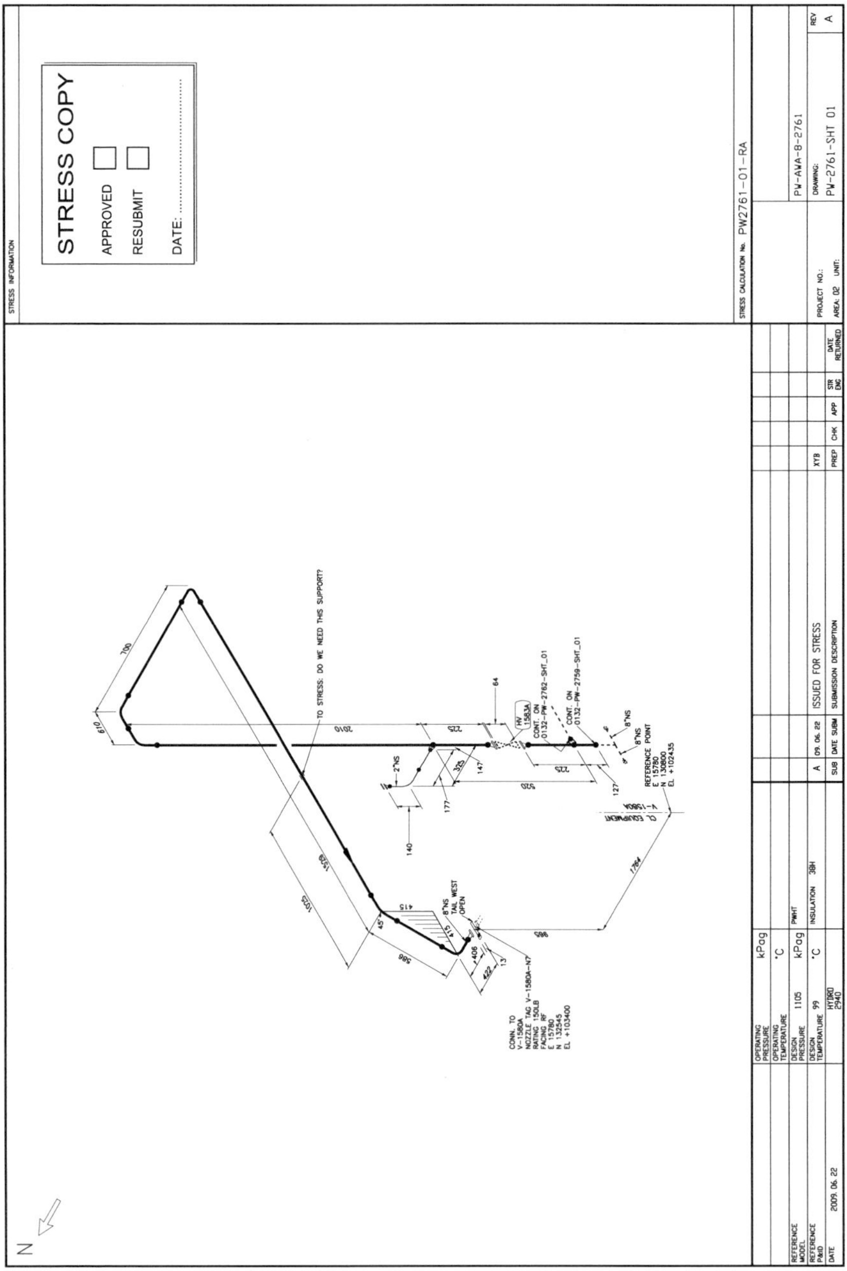

Figure 2–18 *Stress Isometric.*

  - "Yellow-off" on the control P&IDs. This serves as a visual aid to indicate stress progress of lines issued for approval.
  - If more than one person is involved in a system, then the piping designer "yellows-off" their portion and indicates a demarcation, and initials on their side of the demarcation.
  - Ceases work on the piping system while it is being reviewed by the stress engineer.
- Note 2-2—piping lead:
  - Fills-in the stress log.
  - Files the piping copy in numerical order in the "SUBMITTED TO STRESS" book.
  - Issues the stress copy to the stress engineer.
  - For resubmitted stress isometrics, the previous revision returned from stress (in the "RETURNED FROM STRESS" book) is stamped "SUPERSEDED."
- Note 2-3—stress engineer:
  - Imports the importation file into the stress software.
  - Discusses piping configuration issues with the designer. Resolutions are noted on the stress isometric by the stress engineer.
  - Marks-up configuration changes, loads, movements, and tag numbers of pipe supports, anchors, guides, etc.
  - Signs and dates, and marks "APPROVED AS NOTED" or "RESUBMIT" box.
  - Makes a copy for his/her records and returns marked-up isometric to the piping lead.
- Note 2-4—piping lead:
  - Fills-in the stress log spreadsheet.
  - Makes a working copy for the structural lead and the piping designer (stamped "WORKING COPY"), and forwards these copies to them.
  - Files the original into the "RETURNED FROM STRESS" book.
  - Stamps "SUPERSEDED" on the piping copy in the "SUBMITTED TO STRESS" book.

    I recommend removing these from the stress book and storing them separately. In this way you can easily identify which stress isos are still with the stress group.

- Note 2-5—piping designer:
  - Updates the model as per the stress comments.
  - Discusses design issues with the stress engineer. Small changes are marked-up on the master returned copy and initialled by the stress engineer. For exmaple, moving an anchor or a support.
  - Resubmits a revised stress isometric and stress software importation file, if requested.
  - For "APPROVED AS NOTED" lines only, "greens-off" on the control P&IDs over the yellow highlight. This serves as a visual aid to indicate that the piping configurations have been approved.
  - Discusses "V" and "M" designated lines with the stress engineer to get a verbal okay, and provides stress isometrics as requested. Not all lines requiring computer analysis are identified from the minimum requirements and are identified later during design development.
- Note 2-6—piping designer:
  - If it is required to resubmit a revised stress isometric of a greened-off line, then a red cloud is placed around the line on the P&ID with the comment "RESUBMITTED."
  - "Greens-off" the red cloud after re-approval.
- Note 2-7—piping lead:
  - Issues a set of piping arrangement and construction isometric drawings to the stress engineer for the final review at the drawing checking stage.
- Note 2-8—stress engineer:
  - Confirms that stress comments have been incorporated by piping, and reviews lines that were not previously reviewed ("V" and "M" categories).
  - Marks-up any comments.
  - Returns package to the piping lead.

If all of this sounds like a lot of work, let me assure you, it is. But finding out later in the project that lines have been missed can cause very costly delays and rework. There are no easy answers to this.

I worked on a project where we tried importing and exporting from the 3-D models to the stress software and back to the 3-D models. Compatible 3-D and stress software have the capability for import and export, but the procedure for managing the electronic transfer of the converted files through project folders proved to be cumbersome. Lost files, misplaced files and incorrectly named files made keeping track of the progress and getting accurate status reports very difficult and very time consuming to manage.

The immediately visual sense that you get from reviewing the control sets of P&IDs and stress isometric books and the stress log is very valuable. However, you do have to keep the P&IDs up to date with the latest revision issues. Hopefully you will not be receiving frequent changes to the P&IDs after IFE, but if you are, the upside is that you have a great tool for estimating the impact of those changes, and to present an argument against making the changes when you are well into the design. The stick file procedure and this procedure compliment each other. Comments on the stick files, in particular changes to optimize line sizes, may prove to have been marked-up too late when compared with the progress on the stress control set of P&IDs.

You may find that your stress engineer wants to review everything by computer analysis and wants a stress isometric for all lines. Many stress engineers feel that this is the best way to ensure that everything is covered. There is merit to this when you consider that for V and M designated lines the stress engineer will need time to become familiar with the layout by reviewing the piping arrangements and isometrics. They may make dimensional mistakes when adding/subtracting coordinates and elevations. I urge you not to skip the initial review stage because of the focus, as discussed earlier, this will bring to the piping effort. Instead of the V and M designations, leave blanks in the LDT until these lines are stressed. Be aware that creating stress isometrics for all lines will add hours to the piping design schedule, but may well reduce hours for the stress engineering effort.

Finally, it is a good idea for the stress engineers to have access to do their own model walkthroughs to aid them in solving stress problems. However, the job of the stress engineer is to explain an issue to the designer and provide rerouting suggestions, not to redesign the piping for the designer. The designer must work with the stress engineer to come up with an adequately flexible system while taking into consideration the other layout requirements, such as process, clearances, maintenance, and constructability.

2.6 Model Reviews

3-D model reviews are the electronic equivalent of scale plastic model reviews that were very popular in the 1970s and 1980s. Both are a big improvement over the older method of reviewing piping arrangement drawings alone. It takes a lot of time and effort for those unfamiliar with a layout to build a 3-D vision in their mind's eye solely from the drawings, and a lot of layout issues, particularly some of the detail surrounding maintenance and operability, e.g., valve handwheel elevations, would often make it all the way through to construction undetected. One of the first things the operations' personnel used to do after a turnover was to start compiling a wish list of retrofits and design changes.

Models greatly improve the reviewers' abilities to visualize the proposed layouts and lead to a better design and fewer surprises. But, there is a downside. Now everyone thinks that they are a piping designer, and if you are not careful the reviewers will go beyond the intent of the review and request changes based on personal preference. This is particularly true when the designer is not there to explain his/her decisions, and can lead to obvious frustration and unnecessary extra hours. You need ground rules and a strong facilitator to keep things on track in the model reviews:

- Always have an internal model review prior to a review with the client. You do not want to be discussing the design in front of the client.
- Know the parameters and level of design that you are going to discuss. What do you want to achieve in this meeting? For instance, in the initial stages the focus will be of a high level nature, on such things as construction, equipment spacing, and elevation clearances for egress/ingress and maintenance, building sizes, platform and ladder requirements, and future spacing requirements. Getting into discussions about a pressure gauge location will bog down the discussions.
- The project engineer should always be present. You will undoubtedly enter into discussions that will go beyond the project scope of work and outside of the client's own specifications. I've sat in on model reviews where the client's representative wanted platform access to all PSVs, something which was not in his/her own specifications, and staircases instead of ladders. Requests of this nature cannot be acted upon until it has been agreed that this is a scope change with the associated added costs.
- Keep personal preferences out of the discussions, or at least try to minimize them. If you gave two different designers the same area, each would come up with a different layout. As long as they both meet the requirements, one cannot be

considered better that the other. You must adopt the adage "If it isn't broken, don't fix it."

- Do not revisit decisions already made unless new information requires that you do so. The spacing of a set of pumps that come in larger than expected needs to be revisited. The spacing of a set of pumps that come in smaller than expected does not.
- Identify who should be in the model reviews. For instance, if constructability and maintainability are to be discussed make sure that representatives of the construction management team and the client's maintenance group are invited.
- Prior to the client model review, verify that the models to be reviewed are converted to viewing format, that the files are deposited into the correct folders for retrieval and that the conference room computer set-up is in working order. There are very few things in life that are more embarrassing than not being able to access the models when the client is present and you have to call CAD support to correct avoidable issues before the model review meeting can begin.

Most engineering companies will have a model review procedure and a model review matrix based on scheduled reviews at the 30%, 60% and 90% stages of design (note that these are not percentages of project completion). These help to establish the priorities during design and bring focus to the goals of the reviews.

2.6.1 Model Review Procedure

A common model review procedure includes the following:

A statement of the objectives:

- The objectives of the model review is to verify that the current stage of the design meets the minimum project requirements for operability, maintainability, constructability, and functionality, and that it reflects every disciplines' input to the design to date.

The responsibilities that lie with the primary organizers of the model reviews are below.

The project engineer:

- Decides the required attendees, and schedules and manages the formal model reviews.
- Ensures that the design progress meets the requirements for the review as defined in the model review matrix.

- Prepares and issues an agenda for the review.
- Coordinates the discussions and documents resolutions and action items.
- Ensures proper filing of all documentation.
- Distributes the model review comments to the attendees.

The piping lead:

- Coordinates with CAD support the assembly of the 3-D models required for the review.
- Operates or designates a team member to operate the computer "walk through" during the review.
- Ensures that an interference check has been completed prior to the review.
- Ensure that a "snapshot in time" of the 3-D model files used for the review is archived for future reference and record keeping after each review.
- Captures all comments/action items in the software package.
- Maintains a hard copy binder and logs and distributes the action items to the designers.

2.6.1.1 Instruction

- Formal reviews are to take place in three intervals during the development of the IFC 3-D models, defined as 30%, 60% and 90% model reviews. These model reviews are major checkpoints and milestones in the engineering design process. Deliverables required from each discipline to achieve each of these checkpoints are defined in the model review matrix. Typically a model review will not proceed until such time that all of the deliverables required for the specific review have been incorporated. When deliverables are missing for a particular review, the model review may only proceed provided that the associated risks and mitigations have been identified and approved by the project manager.

2.6.1.2 Review Procedures

- Reviewing of the model is a joint responsibility of the piping lead, the designer/drafter, the project engineer, and other related disciplines.

- All comments shall be captured on model screen shots with an assigned action item number. See Figure 2–19 Model Review Comment.
- All proposed changes to the design shall be categorized relative to cost and schedule impacts. This will set the priority for addressing the comments.
- The comments that are not the subject of a scope change shall be incorporated into the working model as a top priority.
- The comments that are the subject of a scope change shall be incorporated into the working model once the scope change has been approved.
- The action items shall be reviewed during the next internal review for completeness, and during the next formal review with the client.
- Regular internal 3-D model reviews, which may not be shown on the project schedule, should take place at time intervals as deemed necessary by the project engineer or piping lead, in conjunction with the other discipline leads. These reviews are considered as routine coordination and part of the design development process.

2.6.2 Model Review Matrix

The model review matrix will outline the preparedness required by each discipline at each stage of review and the items to be reviewed at that stage. As this book is about piping design, we will limit the matrix to the piping aspects. Table 2–4 is an example of a model review matrix.

2.7 Checking

Checking, without question, is one of the most important and necessary functions performed by the piping group. There was a time when the checker's job was to check absolutely everything that went into the drawings, down to the last nut and bolt, so to speak. Today, the process of checking piping arrangements and isometrics has changed, and many aspects of checking are dealt with throughout the life of the project. However, the checker is still the last chance to catch mistakes and omissions before the drawings are issued for construction.

The piping and equipment layouts are reviewed and approved by the client and all the disciplines during the model reviews. Therefore the piping arrangement and isometric checks are not intended to be a

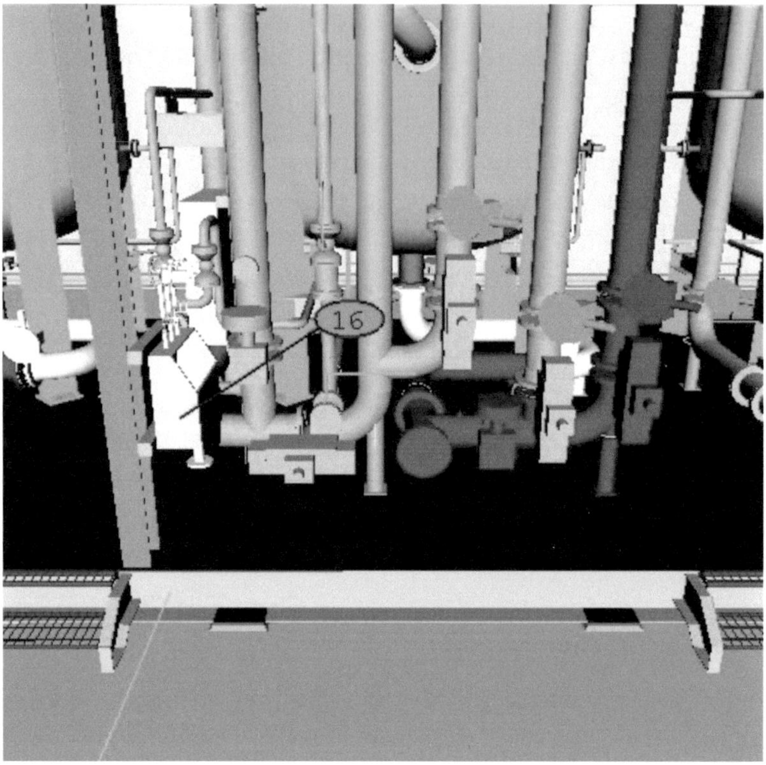

Comment 16

Status: New
Action: Turn sample boxes to front of column.
Date Created: 2008/10/17 19:44:59

20217.298mm, 134592.000mm, 101452.513mm, 20118.300mm, 134048.000mm, 101114.000mm, 20363 300mm, 134592 000mm, 101799.000mm

Figure 2–19 *Model Review Comment.*

Table 2–4 Model Review Matrix

Required By Review Stage	30% Review	60% Review	90% Review
Process Design			
Process Studies	Complete	—	—
PFDs & Mass Balance	IFE	IFC	—
P&IDs	IFA	IFE	IFC
HAZOP	—	Complete	—
Line Sizing	Critical lines complete.	Complete	—
Line Designation Table	IFA	IFE	IFC
Piping Specifications	Approved	—	—
Vendor Info.			
Valves	Preliminary data for NPS 6 and above.	Approved data for NPS 4 and above.	Certified data for all NPS sizes.
Specialty Items	Preliminary data for NPS 6 and above.	Approved data for NPS 4 and above.	Certified data for all NPS sizes.
Equipment	Preliminary	Approved	Certified
Instruments	Preliminary	Approved	Certified
Piping Design			
Plot Plan	IFA	IFE	IFC
Piping Specifications	Calc wall complete.	—	—
Equipment Layout	Major equipment modeled to preliminary information.	Equipment locations finalized and modeled to approved information.	Equipment modeled to certified information.
Piping Stress Analysis	NPS 6 and above critical lines stressed. Other lines under review.	All NPS critical lines stressed. Other lines under review.	NPS 4 and above lines stressed. Other lines under review.

Table 2–4 Model Review Matrix (cont'd)

Required By Review Stage	30% Review	60% Review	90% Review
Piping Design (continued)			
Piping Design	NPS 6 and above lines are modeled and submitted to stress. (See Note 1.)	NPS 4 and above lines modeled and submitted to stress.	All NPS line sizes modeled and submitted to stress. Field welds and hydrotest vents and drains modeled.
Supports	All major temporary steel modeled.	All temporary steel modeled.	—
Tracing Manifolds/Utility Stations/Eye Wash/Safety Showers	Not Started.	Preliminary locations modeled.	Firm locations modeled.
Review Objectives			
	Escape routes, roads, paved areas and drainage.	30% model review comments.	60% model review comments.
	Maintainability, e.g., davits, monorails and cranage, access between equipment, width and height requirements under racks.	Maintainability. Pay particular attention to removal clearances, e.g., exchanger bundle removals, O/H cranes above split casings, etc.	—
	Operability, e.g., platform access.	Operability, e.g., valve heights, chain operators and rising stem requirements.	—
	Constructability, e.g., road access, crane reaches.	Constructability, e.g., field erected piping, module installation.	Field weld and hydrotest vent and drain locations.

Table 2–4 Model Review Matrix (cont'd)

Required By Review Stage	30% Review	60% Review	90% Review
Review Objectives (continued)			
	Future piperack requirements, e.g., widths, number of levels.	—	—
	Main structures.	Minor structures.	Fireproofing of structures.
	Preliminary identification of platforms, ladders, walkways and staircases.	Platforms, ladders, walkways and staircases. Davits, monorails and cranage.	—
	Routing of NPS 6 and above critical lines.	Routing of all NPS critical lines and other lines as required.	Routing of all NPS line sizes as required.
	Pump piping.	Pump piping.	—
	Modularization vs. field erected piping.	Sample Points identified	Tracing Manifold locations.
	—	Utility Station locations.	—
	—	Eye Wash and Safety Shower locations.	—
	Inline instruments and piping configurations.	Inline instruments and piping configurations, and vessel mounted instruments (if available).	Minor instrument locations, e.g., PI and TI, and vessel mounted instruments.

Table 2–4 Model Review Matrix (cont'd)

Ground Rules
1. The cut-off point of NPS 6 and above is a judgment call. Some areas, racks for instance, may require that NPS 2 and above be modeled at this stage. 2. All model review comments are to be addressed immediately following the model review. 3. No changes to approved or frozen items are allowed without Project Engineering approval. 4. The following guidelines have been established for initiating a change. Personal preference is not an acceptable reason: - Design does not meet the design criteria. - Design does not meet accepted design practice. - Design does not meet plant operating standards or client standards for operability, maintainability, fire protection, safe isolation of equipment, or safety requirements. - Design does not meet "What-if" review recommendations.

check of the overall layout or individual pipe routings, but rather a check to ensure that:

- All items and process requirements shown on the P&IDs have been incorporated.
- The correct piping classes have been used.
- The correct standards have been referred to.
- The stress requirements have been incorporated.

Nonetheless, checkers must still use their experience and judgement and be on the lookout for poor piping design. For instance, model reviewers may review the locations of PSVs, but fail to spot a pocket in the relief line. They may review a steam system, but not notice that drip pots and steam traps have not been provided at all low points. They may review exchanger layouts and bundle pulling requirements, and overlook the need for break out flanges.

It is not required to check the following:

- In theory the material descriptions, wall thicknesses (with the exception of calc. wall) and ratings of piping components in the Bills of Material (BOM) do not need to be checked as all of the components are selected from the piping classes which are pre-built and checked at the beginning of the project. In practice I recommend that the checker initially check everything until they have built up a comfort level and can have confidence that the piping class databases are consistently correct. All choices within the piping classes, all out-of-spec components, specialty items and the selected piping classes as per the P&IDs and LDTs must be checked regardless.

- Checking is not required of either the overall or the fitting-to-fitting dimensions on the isometrics. The dimensions of the components are built to the ANSI standards, and provided that the key coordinates and elevations are correct in the model, the dimensions associated to the routing can be taken as correct also. For instance, let us assume that there is a line running from a vessel nozzle down to grade, through a control valve, back up to a stringer elevation, then horizontally across the stringer and dropping into a rack header. Once the coordinates and elevations of the vessel nozzle, the piperack steel and the header location have been verified in the model, the checker will only need to check: the BOP of the horizontal run of the line, the face-to-face dimension and centerline elevation of the control valve, and the clearance for the removal of the actuator.
- Interferences do not need to be checked. Due to the ability to run clash check reports, it is unnecessary to check clearances. That is with one exception; interferences that can be caused by thermal growth have to be investigated.

In short, the items that need checking are those that are decided and input by the designers. All the decisions left to the designer must be checked. It is still a considerable list, as you will see.

The checkers will require access to the live 3-D models and viewing software to actively work with the models to ID tags, line numbers, dimensions, and elevations. Prior to and during the checking of equipment (vendor information) and piping (piping arrangements and isometrics) the following must be observed:

- Models are to be frozen. Designing is not allowed during the checking period.
- After checking, access will be allowed to the designer for backdrafting. Manual edits of the drawings must not be allowed. All changes are to be made in the models and the drawings regenerated for backchecking.
- Once the backchecking is complete, models must be frozen for IFC, and remain so until a revision is required.

The following sections detail the different checking procedures that are required:

2.7.1 Clash Check Reporting Procedure

Under your instruction, the CAD support group will set-up a project directory for the storing of clash reports. They will also help you establish the parameters of the clash reports. For instance, shoes resting on steel may be reported as clashes unless they are automatically removed. Shoes resting on steel can be filtered from the clash reports by setting the tolerances for reporting. Time will be saved by removing the many acceptable clashes from the reports automatically, because these will not have to be cleared individually.

The area leads, or a designated team member, will run periodic clash check reports for each design area during the designing phase, and a final report just before the checking starts. The designers are notified when this has been done. The clash check reports will be stored in the project directory and in a hard copy format for reference in a clash check reports binder(s) in the drafting area.

The six types of clash checks to be run are:

- Piping to Piping (P to P).
- Piping to Equipment (P to E).
- Piping to Steel (P to S).
- Piping to Electrical (P to El.).
- Piping to Building (P to B).
- Equipment to Steel (E to S).

2.7.1.1 Actions

- The clash report will have the following information:
    - Clash report number.
    - Type of clash report (e.g., P to P).
    - Models included in this clash report.
    - Name of the person who ran the clash report.
    - Date of this clash report.
    - Name of the person who addressed this clash report.
    - Date the clash report was addressed.
    - Screen shots of the clashes.
    - Comments
- The piping designers are to refer to the clash check reports to identify clashes. Clashes are to be resolved either by piping

modifications or by negotiation with the structural or electrical designers who will initiate their own changes to resolve the clash.

- When the clashes are resolved the piping designer must run a new clash check report.

2.7.2 Equipment Checking Procedure

2.7.2.1 Purpose

- To ensure the dimensional accuracy and correct location of all vendor equipment built in the equipment models prior to piping arrangement and isometric production and checking.

2.7.2.2 Procedure

- Do not use the piping master vendor drawing from the piping files for checking. Obtain a checking copy of the latest vendor drawing from document control.
- This drawing should be preferably status "Certified" but not less than status "Approved."
- The vendor drawing and any comments should be the same as the piping master copy in the piping files, but check that there are no major unincorporated comments when compared to piping master copy. If major comments are still unresolved, do not continue the check. Inform the piping lead.
- Stamp the vendor drawing "Check Print" and date.
- Obtain the current equipment nozzle report. Stamp "Check Print" and date. Refer to Figure 2–20 Nozzle Report.
- Obtain a screen print of the equipment dialogue box and stamp "Check Print." Refer to Figure 2–21 Equipment Dialogue Box.
- ID the centerline coordinates, elevation and orientation from the equipment model and write them on the vendor check print.
- ID the underside elevation of base plates and the TOC or TOS of the foundation and write them on the vendor check print.
- Compare the coordinates and elevation with those noted on the master stick file equipment location plans.

Friday, July 10, 2009 *Nozzle Schedule Report*

Equipment Tag	Nozzle Tag	Description	Size	Rating	Projection (overall Length)	North	East	Elevation	Direction
V-1580A									
	V-1580A-N10A	NOZZLE 150LB RF	2	150LB	133 mm	131638 mm	14329 mm	104467 mm	W 30.00 N
	V-1580A-N1	NOZZLE 150LB RF	8	150LB	1300 mm	130800 mm	15780 mm	106783 mm	U
	V-1580A-N2	NOZZLE 150LB RF	8	150LB	914 mm	131562 mm	15780 mm	101521 mm	D
	V-1580A-N4	NOZZLE 150LB RF	3	150LB	1559 mm	131902 mm	14678 mm	104590 mm	N 45.00 W
	V-1580A-N5	NOZZLE 150LB RF	3	150LB	799 mm	130800 mm	15780 mm	101636 mm	D
	V-1580A-N6	NOZZLE 150LB RF	2	150LB	1300 mm	130800 mm	16136 mm	106783 mm	U
	V-1580A-N7	NOZZLE 150LB RF	8	150LB	202 mm	132545 mm	15780 mm	103400 mm	N
	V-1580A-N8A	NOZZLE 150LB RF	2	150LB	133 mm	130800 mm	14104 mm	102638 mm	W
	V-1580A-N8B	NOZZLE 150LB RF	2	150LB	133 mm	130800 mm	17456 mm	102638 mm	E
	V-1580A-N9	NOZZLE 150LB RF	3	150LB	133 mm	129615 mm	16965 mm	105026 mm	E 45.00 S
	V-1580A-N10B	NOZZLE 150LB RF	2	150LB	133 mm	131638 mm	14329 mm	102562 mm	W 30.00 N
	V-1580A-M1A	AXIAL ROOF MANWAY	24	150LB	1300 mm	130800 mm	15018 mm	106831 mm	U
	V-1580A-M1B	API STANDARD 650 RADIAL SHELL MANWAY	24	150LB	257 mm	129061 mm	15314 mm	103045 mm	S 15.00 W

1

Figure 2–20 *Nozzle Report.*

- Yellow off the equipment location plan coordinates and elevation, or mark up the equipment location plan with any discrepancies after consultation with the piping lead.
- Check the nozzle report and the equipment dialogue box against the vendor drawing and yellow off all documents.
- Mark up any discrepancies on the nozzle report (at this stage it may be too late to revise the physical equipment) point out discrepancies to the piping lead who will request the changes to be made in the model if necessary.
- Check the dimensions on the vendor drawing against the equipment model carefully. The reference points used when building the equipment model may differ from the dimensions provided on the vendor drawing. For instance, horizontal drums are often built using seam-to-seam, whereas the vendor drawing may be dimensioned tangent-to-tangent.
- Gather together all checked documents and make up a file folder marked "Check Prints (Equipment No)." If there are no discrepancies in the checking package, place the folder in the piping equipment file behind the master copy of the vendor

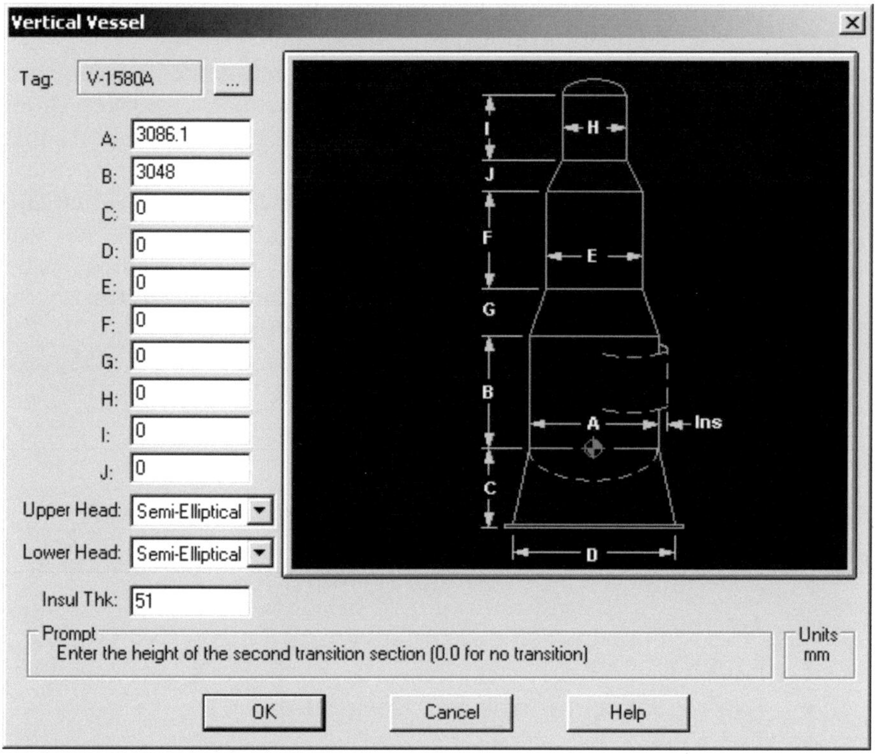

Figure 2–21 *Equipment Dialogue Box.*

drawings. Do not file before pointing out any discrepancies to the piping lead.

- When checking is complete and changes, if required, have been made, the piping lead will instruct the designer to freeze the equipment model.

2.7.3 Piping Arrangement and Isometric Checking

There was a time when piping arrangements were fully detailed, dimensioned, and sectioned, primarily for one simple reason; the person drawing the isometrics required fully detailed and dimensioned piping arrangements to work with. Due to the ability to automatically generate annotated and dimensioned isometrics from the model, fully detailed and dimensioned general arrangements are now largely unnecessary. A secondary reason was for the benefit of the piping erectors. I say secondary because once you have isometrics to go by, the full dimensioning and sectioning on the piping arrangements takes on a lesser importance.

Piping arrangements are still an important reference document for construction, a document that shows the relationships between pipes, equipment, supports, and platforms and in this context it is only necessary to provide sufficient information to the erectors. Now that 3-D models are available for reference and isometrics are automatically generated, piping arrangements should be minimally detailed and dimensioned, and as far as possible, be limited to plans only. Partial sections and details should only be required when a congested area or some vertical detail cannot be clarified in any other way.

The following is the information commonly contained within the plans:

- Plant North arrow.
- Key plan.
- General notes.
- Matchlines, coordinates, and continuation references.
- Area limits.
- Equipment tags.
- Equipment coordinates and elevations.
- Nozzle tags.
- Sliding and fixed saddles.
- Identify module tag numbers and limits.
- Datum point and plant coordinates located on the NW corner steel center-lines of the piperack or equipment modules.
- Shipping notes (piperack and equipment modules only).
- Pipe support and equipment supporting steel numbers and coordinates.
- Building tag numbers.
- Line numbers.
- Flow arrows.
- Reducer identification at equipment nozzles (inline identifications are not required).
- Miscellaneous line information, e.g., spec. breaks and tie-in numbers.
- Instrument tags.
- Location of shoes, anchors, guides, etc.

- Bottom Of Pipe (BOP) elevations.
- Centerline work point elevations and call-outs for trimmed shoes and shim heights on slopped lines.
- Miscellaneous information, e.g., cable trays and safety shower and eye wash tag numbers.
- Critical dimensions only. "Critical dimensions" are those that tie-down piping connections for the fabricators, the locations of which are essential in order to ensure match-up of a continuation, e.g., module to module or module to field erected. They are also dimensions from a reference point to vertical lines passing between upper and lower piping arrangements (these dimensions appear on the lower piping arrangement only), and reference dimensions required to assist the erectors, e.g., centerline of column to centerline of pipe.

As much as is possible should be checked within the models themselves. Keeping in mind that the clash reports and equipment checks are done, the checker can focus primarily on checking that the piping in the models matches the latest P&IDs and LDTs and that the required information is included in the drawings. In other words, when the models are checked to ensure accuracy, the information that flows from them and appears on the drawings must be accurate also. The drawing checks are only required to ensure that the content on the drawings is complete.

It used to be a common practice to check the isometrics at a later date after the piping arrangements were completed and sufficient time had elapsed to draw these isometrics. As all information now comes from the model databases, this is no longer the case. The checking of the piping arrangements and the associated isometrics can be scheduled to happen together or separately. My preference is that the piping arrangements and isometrics be checked together. This way all information is before the checker at the same time. The downside to this is that the piping arrangements must be generated in order to do so. Another method is to check the isometrics first, make the changes to the models, and then generate the piping arrangements and check them during the backchecking of the isometrics. The upside to this is that the piping arrangements are not generated until after the models are corrected. The downside is that the checker does not have a complete package to work with until after the initial check. The choice is yours. You may want to try it both ways and see which approach works best for you.

2.7.4 Prerequisites and Checking Procedure

Before piping arrangement and isometric checking begins, the following documents should be complete. It is the responsibility of the checker to obtain and use the most current copy of the:

- Instrument list and data sheets.
- Tie-in list (if appplicable).
- Specialty item list and data sheets.
- Piping Job Notes (PJN).
- Piping classes.
- Piping standards.
- IFC P&IDs.
- IFC Line Designation Tables (LDT).
- 90% model review comments.
- Final stress isometrics.
- Spring support data sheets.
- Structural steel drawings issued for checking.
- Equipment check files.
- Recent clash check report run and cleared.

Prior to commencement of checking, the piping arrangements and isometrics should be put on an IDR. Any comments arising from the IDR should be reviewed by the checker and incorporated into the piping arrangements and isometrics during the backdrafting.

2.7.4.1 Piping Arrangement Drawing Content Check List (must be established on the drawing)

- Plant North arrow (up or to the left) in top left hand corner of drawing.
- Key plan in top right hand corner of the drawing.
- Title block:
  - Drawing number.
  - Drawing title.
  - Project name and number.
  - Revision number.
  - Revision description and date.

  - Drawing scale.
  - Reference drawings.
- Matchline coordinates and references to adjoining drawings.
- Area limit coordinates.
- Continuations to underground drawings.
- Equipment tag numbers, coordinates, elevations and nozzle tags (note: these are already checked. Verify the information on the drawing).
- Flow arrows.
- General notes.
- Call-outs and notes for any temporary shipping supports or strapping, or for any part of a line that is shipped loose (piperack and equipment modules only).
- Critical dimensions.
- Insulation breaks are 500mm back from all termination points of butt welds to allow for later field welding (piperack and equipment modules only).
- Cloud "HOLDS" and create a holds list.

2.7.4.2 Piping Arrangement Design Content Check List (checked in the model and verified on the drawing where applicable)

- Pipe and equipment support numbers, coordinates and TOS elevations.
- Modules are within the width, height and length allowances for shipping requirements.
- Line spacing. Check that line spacing also allows for expansion growth at changes of direction. Refer to Table 2–5 Thermal Expansion Chart.
- Correct call outs for shoes, anchors, guides, etc., per the stress isometrics.
- Bottom of Pipes (BOP) elevations.
- Centerline work point elevations and trimmed shoe and shim heights for sloped lines.
- Call outs for reducers/swages at equipment nozzles: concentric, eccentric (FOB, FOT) with sizes.
- Adequate space provision to swing spades and remove blinds.
- Adequate space provision for valve rising stems.

Table 2–5 Linear Thermal Expansion between 70°F/21°C and Indicated Temperature. (Inches per 100 feet or millimetres per 30.48m)

Temp		Carbon Steel, Low Chrome (thru 3 Cr Mo)		5 Cr Mo thru 9 Cr Mo		Austenitic Stainless Steels	
°F	°C	inches	mm	inches	mm	inches	mm
–50	–46	–0.84	–21	–0.79	–20	–1.24	–32
–25	–32	–0.68	–17	–0.63	–16	–0.98	–25
0	–18	–0.49	–12	–0.46	–12	–0.72	–18
25	–4	–0.32	–8	–0.30	–8	–0.46	–12
50	10	–0.14	–4	–0.13	–3	–0.21	–5
70	**21**	**0.00**	**0.00**	**0.00**	**0.00**	**0.00**	**0.00**
100	38	0.23	6	0.22	6	0.34	9
125	52	0.42	11	0.40	10	0.62	16
150	66	0.61	15	0.58	15	0.90	23
175	80	0.80	20	0.76	19	1.18	30
200	93	0.99	25	0.94	24	1.46	37
225	107	1.21	31	1.13	29	1.75	44
250	121	1.40	36	1.33	34	2.03	52
275	135	1.61	41	1.52	39	2.32	59
300	149	1.82	46	1.71	43	2.61	66
325	163	2.04	52	1.90	48	2.90	74
350	177	2.26	57	2.10	53	3.20	81
375	191	2.48	63	2.30	58	3.50	89
400	204	2.70	69	2.50	64	3.80	97
425	219	2.93	74	2.72	69	4.10	104
450	232	3.16	80	2.93	74	4.41	112
475	246	3.39	86	3.14	80	4.71	120
500	260	3.62	92	3.35	85	5.01	127
525	274	3.86	98	3.58	91	5.31	135

Table 2–5 Linear Thermal Expansion between 70°F/21°C and Indicated Temperature. (Inches per 100 feet or millimetres per 30.48m) (cont'd)

Temp		Carbon Steel, Low Chrome (thru 3 Cr Mo)		5 Cr Mo thru 9 Cr Mo		Austenitic Stainless Steels	
°F	°C	inches	mm	inches	mm	inches	mm
550	288	4.11	104	3.80	97	5.62	143
575	302	4.35	110	4.02	102	5.93	151
600	316	4.60	117	4.24	108	6.24	159
625	330	4.86	123	4.47	114	6.55	166
650	343	5.11	130	4.69	119	6.87	175

Identify the pipe material and the design temperature (use the higher number for in-between temperatures), then apply the following formula to calculate the expansion:

$$\frac{\text{length of pipe in feet}}{100} \times \text{exp. per 100 feet} = \text{exp. in inches}$$

or:

$$\frac{\text{length of pipe in meters}}{30.48} \times \text{exp. per 30.48 meters} = \text{exp. in millimeters}$$

- Left or right hand instruments correctly orientated, e.g., ball pattern control valves.
- Personnel protection insulation limits.
- Drop zones.
- ID the TOS to underside of shoe spring heights, or underside of steel to centerline of pipe of hanger rods and hanger springs.
- Sliding/fixed saddle locations.
- Vendor package tag numbers and limits.
- Module tag numbers and limits.
- Datum point and plant co-ordinates of piperack and equipment modules.
- Building tag numbers.
- Line numbers.
- Spec. breaks.
- Tie-in numbers.
- Instrument tag numbers.
- Cable trays shown.

- Utility station tag numbers.
- Safety shower and eye wash station tag numbers.
- Sample cooler tag numbers.

There is software now available that shows great promise and will provide for automatic piping arrangement generation with some minor manual editing. The drafting time that is currently expended on manually annotating the piping arrangements is reduced, but the checking time is not reduced. Whether the annotation and dimensioning be manually placed or automatically generated does not change the fact that the design aspects in the model and the information on the drawings must be checked.

Because there are limitations to the automation that can be achieved by these software, much like the adjustments in expectations we all had to make with automatically generated isometrics, compromise is required here too. As with the isometrics, there is the ability to select parameters, and while you may not achieve 100% of everything you desire, it is a question of function over form that saves drafting hours. I believe the key is to ensure that the end users can obtain all of the needed information by reading the piping arrangements and isometrics in conjunction with one another.

2.7.4.3 Isometric Drawing Content Check List (must be established on the drawing)

- Plant North arrow up and to the left (preferred in industry) or up and to the right per company standard.
- Title block:
    - Drawing and sheet number.
    - Project name and number.
    - Revision number.
    - Revision description and date.
    - Line number.
    - Area number.
    - Model number reference.
    - P&ID reference.
    - NDE and PWHT requirements.
    - Painting, insulation, and tracing requirements.
    - Code, e.g., B31.3.

- Check the operating and design conditions against the master stick file LDT. In theory, these will be correct because they are imported from the native spreadsheet, but do not assume this to be true. In reality the updating of the LDT native spreadsheet often trails behind the plotting of the isometrics for checking.
- Isometric continuation references.
- Equipment tag numbers, coordinates, elevations and nozzle tags (note: these are already checked. Verify the information on the drawing).
- Flow arrows
- Isometrics without a piece of equipment in them, i.e., a nozzle, require a reference to a column centerline.
- Notes for base guides are correct, e.g., "ALLOW N-S or E-W MOVEMENT."
- Cloud "HOLDS" and create a holds list.

2.7.4.4 Isometric Design Content Check List (checked in the model and verified on the drawing where applicable)

- Components in the BOM (note: do not check quantities).
- Correct piping class used.
- The correct calculated wall thickness is called-out.
- Branch lines checked against the branch chart, and for vapour and liquid traps. It may seem unnecessary to check the branch fittings given that the choices available are pre-determined and limited per the piping class database. However, I have seen reducers added to branch lines on P&ID's and designers who have followed this exactly, placing a straight tee and reducer when a reducing tee should have been used.
- Call outs for shoes, anchors, guides, etc., are per the standards, and the locations and the piping configurations are per the latest stress isometrics.
- Pump suction piping checked for eccentric (FOT) reducers and 5 to 10 straight pipe diameters prior to the suction nozzle.
- Hydrostatic vents and drains have been provided at all high and low points, and are per standards and piping class, e.g., double block.
- Instrument connections are per standards and piping class, e.g., flanged thermowells versus threaded or socket welded.

- Piping configurations for inline instruments conform to the data sheets and standards, e.g., upstream and downstream requirements have been met.
- Orifice flanges are Class 300 minimum, and tap orientations are identified.
- Rotated valves angles are possible; e.g., a valve with a flange pattern of eight bolts rotated to 45 degrees is possible. A valve with a flange pattern of twelve bolts rotated to 45 degrees is not.
- Check valves are positioned in the direction of flow. Note that swing disk type check valves cannot be positioned in the vertical with the flow down.
- Out-of-spec items are identified (e.g., flanges at pumps).
- Spool sizes are within the shipping box size, Mark numbers are noted and field welds and field fit-up welds are logically located, e.g., field welds in the horizontal plane when possible, field fit-up welds between rotating equipment.
- Non-standard bolt lengths on PSV inlets, wafer valves, and bleed rings.
- Shop material and field material split is correct in the BOM.
- Centerline elevations of control valve stations above grade and platforms are at the height required by the standards, i.e., adequate height to ensure drain valve clearance.
- Dimensional checking is to be limited to:
  - Face to face dimensions of in-line instruments.
  - Face to face dimensions of specialty items.
  - Clearance is adequate for the removal of control valve operators.
  - Clearance is adequate for the removal of equipment.
- Specialty items (SP) noted.
- Tie-in numbers noted.
- Platform, floor, and wall penetrations are shown.

A checking file is to be created. Initially this may be compiled by the designer and added to by the checker. This file must contain all of the drawings and information used or created for the check, and is to be stored for future record.

Backdrafting must be done in the models and the drawings regenerated. Never manually edit the automatically generated parts of any drawings using straight 2-D CAD.

After the checks and backdrafting are completed the models are frozen, and prior to IFC a "Yellow Off" of the lines is conducted by the piping lead, or designate, on the master P&IDs. The piping arrangements and isometrics are followed and demarcations are noted to indicate the extent of the lines checked. This is one last quality check to ensure nothing slips through the cracks.

2.8 Manhour Estimating and Manpower Planning

Manhour estimating is the exercise of identifying the activities and estimating the number of hours that each activity will take, in order to arrive at the total number of hours required to complete the project. Manpower planning is the identification of the required skill sets and assigning of the required number of personnel to these activities to support the schedule.

When it comes to the detailed estimate of the piping manhours and planning the piping manpower, it is the potential to overlook the detail that can cause your biggest headaches later on. You must consider all of the activities that you are responsible for and the skill sets you will require in order for your part of the project scope to be accomplished. Make yourself aware of the following before starting:

- Project scope definition: What are you designing and building? What activities are to be included? Read the Design Basis Memorandum (DBM) and the Project Execution Plan (PEP).
- Specifications and standards: Are you using existing company or client specifications and standards, or will you be developing them?
- Deliverables: What are your deliverables? For exmaple, 3-D models, piping arrangements and isometrics?
- Fabrication and construction strategies: For instance, will this project entail modularization?
- Are you responsible for material control and/or stress analysis? It is not unusual for these activities to come under the management of the piping lead.

Prior to starting your manhour estimating and manpower planning, divide the plot plan as described below. It is very likely that your first cut at this will have to be refined as you work on your estimates because the construction execution plan, i.e., module versus non-module and construction sequencing, is firmed-up during the estimating process and the study phase of detailed design.

Dividing the plot plan will give you a feel for the complexity of the facility and will be particularly useful for the manpower planning. For smaller projects the first bullet below may be ignored and where I refer to "plant area" in the following bullets, read "plot plan".

For larger projects compile your estimates according to each plant area. The division of the plot plan is as follows:

- Divide the plot plan into the general geographical plant areas, e.g., tank farm area, hydro-treating area, water treatment area.
- Divide each plant area into CWP boundaries according to field erected and module packages, and the construction schedule.
- Divide each CWP into model areas or designer areas. You will require a sufficient number of models in each CWP in order to assign the work to enough individual designers to meet the schedule. Flexibility is the consideration; flexibility to parachute in another designer if needed, and flexibility to check and issue in stages.
- Divide each model area into piping arrangement drawings. Do not overlook that there may be multiple levels requiring separate plans for each.

For a more detailed explanation on plot plan division refer to Chapter 4.

2.8.1 Manhour Estimating

First of all, what is an estimate? It is your best educated guess. An educated guess being the best you believe you can do based on:

- Historical data gathered from similar projects.
- The level of detail/scope provided.

You should always approach your estimating from the point of view that everything will run smoothly. We both know that this will not be the case, however, do not pad your estimate with "what if" scenarios. Come up with a realistic number and leave it to the project management team to put in the contingencies.

Create a manhour estimate spreadsheet where you list each of the deliverables and the number of each deliverable, each of the activities and the hours associated with each activity, and the totals. This will be an accurate record of the basis for your estimate and what you agreed to. It will aid you greatly later when you have to submit estimates for trends and scope changes.

The below list of deliverables and activities for a 3-D project is not a definitive list, and you may well identify items to be added, items that do not apply or items that you would combine into one for your own situation. My goal here is to point out that a good estimate requires thought and preparation so that you do not overlook anything.

- Number of P&IDs and the drafting hours required.
- Piping design hours.
- Number of piping arrangements and hours for generation.
- Number of isometrics and hours for generation.
- Number of heat tracing drawings and hours for the tracing design.
- Checking hours.
- Administration hours.
- CAD support hours.
- Material control hours.
- Stress analysis hours.
- Hours for the development of piping classes and piping standards.
- Hours for the inputting and checking of the piping classes and standards into the databases.
- Hours for as-builting and project close-out.
- Supervision hours.
- On revamp projects: hours for gathering of existing information/drawings, field trips for meetings, gathering of field information, and hanging of tie-in tags.

Search out historical data, as this will aid greatly to make your estimate as accurate as possible and to back-up your numbers.

Historical data, compiled at the end of a project should contain the following:

- The type and size of the plant.
- The design and drafting software used, i.e., 2-D CAD, 3-D CAD.
- A list of the deliverables, i.e., piping arrangements and isometrics.
- The manhours expended for each activity, e.g., design, checking, supervision, stress analysis, CAD support.

- The manpower loading for the duration of the project, e.g., forty designers for two years.

There are four basic methods used to estimate the piping design hours for a project. I recommend that you do at least two estimates for your project using two different methods for comparison.

Methods of estimating the piping design hours are:

- Equipment count.
- Nozzle count.
- Line count.
- Drawing count.

The estimate for the piping design hours is used to estimate some of the other activities on a percentage basis, e.g., CAD support. Remaining activities are added as a function of time, e.g., supervision.

The following is an example of estimated hours. These figures come from numbers used for 3-D grass root refineries and chemical plants. This is why historical data is so important, as different types of projects may or may not utilize similar numbers. These figures include design development time. For duplicated equipment or trains, estimate the piping design hours for the first at 100% and 50% for the subsequent duplicated equipment or trains. Estimates are made from the low number to the high number within a range, depending on the design complexity.

Use one of the following four methods to estimate the piping design hours:

- Equipment counts obtained from the PFDs and P&IDs. Allow 150 to 400 hours per piece of equipment for piping design hours.
- Nozzle count obtained from the P&IDs. Allow 20 hours per nozzle for piping design hours. Include vents, drains, and instrument connections, etc.
- Line count obtained from the P&IDs and the LDT. Allow 30 hours per line for piping design hours.
- Drawing count: allow 100 to 200 hours per piping arrangement for piping design hours.

Add the hours for the below activities to get the total estimated hours:

- P&IDs: allow 20 hours per drawing for drafting for new drawings and 5 to 10 hours for development of already created drawings.

- Piping arrangement generation: allow 5 to 8 hours per drawing for drawing generation.
- Isometric count: allow 2.5 isometrics per line and 2 to 3 hours per isometric for drawing generation.
- Tracing design and checking (hot oil, steam, or glycol): 10% to 20% of the piping design hours.
- Checking: 20% to 25% of the piping design hours.
- Administration, e.g., stress isometric coordination. This is a function of the project duration. Allow hours for the duration of the project, usually 40 hours per week per person.
- CAD support: 5% to 10% of the total piping hours.
- Material take-off group: 8% to 12% of the piping design hours.
- Stress analysis: 10% to 15% of the piping design hours.
- Piping class and piping standard drawing development: 1% to 3% of piping design hours dependent on whether you are accepting, modifying, or creating the client's piping classes and standard drawings.
- As-builting and project close-out: 1.5 hours per drawing.
- Supervision: overall piping lead and area leads are a function of the project duration. Allow hours for the duration of the project, usually 40 hours per week per person. Commonly there will be one lead per plant area, but you may have a small area that can be merged with a larger area, or a couple of smaller areas that can be managed under one lead to distribute the responsibilities evenly.

The hours above are based on 3-D CAD however, the differences of the overall piping hour's budget for a project using 3-D CAD, 2-D CAD or manual methods for design and drafting are not as appreciable as some people may believe. This is primarily due to the following reason:

- The time requirements for the planning and administration by the leads and the thought processes required of the designers and checkers that go into finished designs has not been reduced significantly. CAD software primarily reduces drafting time. The bulk of the hours, seventy five to eighty five percent of the piping hours budget has always been, and is still used in the supervision, administration, designing and checking processes, and this does not change much regardless of the tool. Thinking and executing are still the largest parts of the piping effort.

The savings in the number of hours for a 3-D CAD project over a 2-D or manual project are primarily the benefits and efficiencies that are realised by the downstream disciplines:

- The abilities to run reports for bid and purchasing purposes.
- The visualization that the models provide for design reviews internally and externally with the client.
- The visualization that the models provide for construction planning and execution.
- The improved attainable accuracy over manual and 2-D leading to reduced field rework and lower Total Installed Cost (TIC).

The theoretical design, drafting and checking hours saved from the piping budget on a 3-D project are offset as the piping group now have added responsibilities and need added hours that were not previously required.

- The databases require maintenance, and the piping designers are responsible for doing their part.
- There will be downtime due to problems associated with the models.
- Junior designers do not get the level of training and mentoring prior to being involved in the layout of piping as was the case prior to 3-D CAD, which can result in added hours for checking and rework. The training that is provided primarily focuses on the use of the software and not on piping layout and design.
- There is a whole new group of professionals required, the CAD support group, whose time must be paid for by the project, and who are often part of the piping budget. This is true of 2-D also, but not to the same extent.

The bottom line is that while we have saved some piping hours in the areas of designing, drafting and checking using CAD, in other ways we have only modified the methods and order of piping execution, substituted one activity for another and added piping hours in areas of software complexity, database management and CAD support requirements.

The next step in the estimating process, once you have submitted your hours to the project management, is the negotiation. They will have a number in mind factored from the capital cost of the project. If you are close to their number, your budget will most likely be accepted. However, if you are widely apart from their number you will be challenged and a negotiation will begin. When you are challenged, ask educated questions about the basis for the project management's numbers. Do the numbers they are using include all of the activities and/or deliverables that you have budgeted for? You will have to revisit your estimate and rework your numbers, but try not to be intimidated into an agreement that makes you uncomfortable. Be realistic. Your negotiation may be to strike certain activities/deliverables off of your list, or it may be to reduce your hours in an area where initially you may have been overly cautious. Eventually you will reach a number that suits you both.

Estimating is not an exact science and most likely you will get it wrong. Quite frankly it is an area that most leads struggle with. But the good news is that things often balance out; you'll underestimate in one area and overestimate in another. Over a period of time as you gain experience you will also develop a gut feel and your ability to estimate will get better.

Once you have the budgeted hours established you have another task ahead, that of manpower planning to support the schedule.

2.8.2 Manpower Planning

Again, manpower planning is not an exact science, but the taking of certain steps will help you bore down to a plan:

- Divide the plot plan.
- Apportion the hours.
- Load level.
- Staff appropriately.

2.8.2.1 Dividing up the Plot Plan

- Presumably, the plot plan has been divided into the general geographical plant areas, CWP boundaries, models and piping arrangements.

2.8.2.2 Apportioning the Hours

- If you haven't already done so, you have to apportion the total budgeted manhours into each plant area. Break the hours of each plant area down further into each CWP. The manhours for each CWP are further divided into the model/design areas depending upon the complexity of the CWP. I personally use the piping arrangements to use as an indicator. Based on the number of piping arrangements in each CWP, I decide on the hours for each model/design area by giving a little and taking a little depending on the level of complexity.
- Create a manhour spreadsheet: on the left hand side list the plant areas (rows), subdivided by CWP/ model/design areas, and further by design and checking. Across the spreadsheet is the timeline divided by weeks (columns). See Figure 2–22 Manhour Spreadsheet Example.
- By referring to your estimate and the schedule, divide the hours, usually forty per week to give the number of weeks, and plug in the weekly hours for each model/design area. Remember that the issue dates will be the out-the-door dates and you will have to allow time for plotting and sign-off of the drawings. Working backward from the date that the checking must be completed will give you the start date that the designing must begin.

2.8.2.3 Load Levelling

- Once your spreadsheet has been completed, it will become apparent that you have peaks and valleys in the workload. Even the peaks and valleys out by moving model/design areas forward or back by a week or two. e.g., the schedule requires 500 isometrics to be issued per week for a couple of weeks, followed by 100 isometrics per week for a couple of weeks. The logistical difficulties encountered in issuing 500 isometrics per week can be overcome by relaxing the schedule and spreading these out into the coming weeks. Refer to Figure 2–23 Manhour Graph Example. You will have to go back to the project management team and renegotiate the schedule somewhat, but this is the time to do it. Later, when fabrication shop time has been booked, the pressure will be intense to stay on schedule. Not to mention that your abilities as a lead may be questioned if you start slipping from the schedule and have not been proactive in identifying and resolving problem areas.

MANPOWER REQUIREMENTS					PROJECT MANPOWER BY WEEK - STARTS ON MONDAY																									
					2010																									
AREA	ACTIVITY	Name	Est. Hours	Est. M/W	4-Jan	11-Jan	18-Jan	25-Jan	1-Feb	8-Feb	15-Feb	22-Feb	1-Mar	8-Mar	15-Mar	22-Mar	29-Mar	5-Apr	12-Apr	19-Apr	26-Apr	3-May	10-May	17-May	24-May	31-May	7-Jun	14-Jun	21-Jun	28-Jun
ALL AREAS																														
	Piping Lead		1040		40	40	40	40	40	40	40	40	40	40	40	40	40	40	40	40	40	40	40	40	40	40	40	40	40	40
	CAD Support		960		16	16	24	24	40	40	48	48	48	48	48	48	48	48	48	48	48	48	48	48	32	32	24	24	8	8
	Material Control		480		8	8	12	12	20	20	24	24	24	24	24	24	24	24	24	24	24	24	24	24	16	16	12	12	4	4
	Stress Analysis		480		8	8	12	12	20	20	24	24	24	24	24	24	24	24	24	24	24	24	24	24	16	16	12	12	4	4
	Iso Coordination		640												40	40	40	40	40	40	40	40	40	40	40	40	40	40	40	40
Sub-Total All Areas Hours			**3600**		**72.0**	**72.0**	**88.0**	**88.0**	**120.0**	**120.0**	**136.0**	**136.0**	**136.0**	**136.0**	**176.0**	**176.0**	**176.0**	**176.0**	**176.0**	**176.0**	**176.0**	**176.0**	**176.0**	**176.0**	**144.0**	**144.0**	**128.0**	**128.0**	**96.0**	**96.0**
Total Area All Hours/Manweeks			**3600**	**90**	**1.8**	**1.8**	**2.2**	**2.2**	**3.0**	**3.0**	**3.4**	**3.4**	**3.4**	**3.4**	**4.4**	**4.4**	**4.4**	**4.4**	**4.4**	**4.4**	**4.4**	**4.4**	**4.4**	**4.4**	**3.6**	**3.6**	**3.2**	**3.2**	**2.4**	**2.4**
PLANT AREA 01																														
	Area Lead		1040		40	40	40	40	40	40	40	40	40	40	40	40	40	40	40	40	40	40	40	40	40	40	40	40	40	40
Sub-Total Area 01 Lead Hours			**1040**		**40.0**	**40.0**	**40.0**	**40.0**	**40.0**	**40.0**	**40.0**	**40.0**	**40.0**	**40.0**	**40.0**	**40.0**	**40.0**	**40.0**	**40.0**	**40.0**	**40.0**	**40.0**	**40.0**	**40.0**	**40.0**	**40.0**	**40.0**	**40.0**	**40.0**	**40.0**
CWP Number 01-1 - Model 1	**Designing**	Designer 1	320		40	40	40	40	40	40	40	40																		
CWP Number 01-1 - Model 2	**Designing**	Designer 1	320										40	40	40	40	40	40	40	40										
CWP Number 01-2 - Model 1	**Designing**	Designer 2	320						40	40	40	40	40	40	40	40														
CWP Number 01-2 - Model 2	**Designing**	Designer 2	320														40	40	40	40	40	40	40	40						
Sub-Total Areas 01 Designing Hours			**1280**		**40.0**	**40.0**	**40.0**	**40.0**	**80.0**	**80.0**	**80.0**	**80.0**	**80.0**	**80.0**	**80.0**	**80.0**	**80.0**	**80.0**	**80.0**	**80.0**	**40.0**	**40.0**	**40.0**	**40.0**	**0.0**	**0.0**	**0.0**	**0.0**	**0.0**	**0.0**
CWP Number 01-1 - Models 1 and 2	**Checking**	Checker 1	160																		40	40	40	40						
CWP Number 01-2 - Models 1 and 2	**Checking**	Checker 1	160																						40	40	40	40		
Sub-Total Area 01 Checking Hours			**320**		**0.0**	**0.0**	**0.0**	**0.0**	**0.0**	**0.0**	**0.0**	**0.0**	**0.0**	**0.0**	**0.0**	**0.0**	**0.0**	**0.0**	**0.0**	**0.0**	**40.0**	**40.0**	**40.0**	**40.0**	**40.0**	**40.0**	**40.0**	**40.0**	**0.0**	**0.0**
Total Area 01 Hours/Manweeks			**2640**	**66**	**2.0**	**2.0**	**2.0**	**2.0**	**3.0**	**3.0**	**3.0**	**3.0**	**3.0**	**3.0**	**3.0**	**3.0**	**3.0**	**3.0**	**3.0**	**3.0**	**3.0**	**3.0**	**3.0**	**3.0**	**2.0**	**2.0**	**2.0**	**2.0**	**1.0**	**1.0**
PLANT AREA 02																														
	Area Lead		1040		40	40	40	40	40	40	40	40	40	40	40	40	40	40	40	40	40	40	40	40	40	40	40	40	40	40
Sub-Total Area 02 Lead Hours			**1040**		**40.0**	**40.0**	**40.0**	**40.0**	**40.0**	**40.0**	**40.0**	**40.0**	**40.0**	**40.0**	**40.0**	**40.0**	**40.0**	**40.0**	**40.0**	**40.0**	**40.0**	**40.0**	**40.0**	**40.0**	**40.0**	**40.0**	**40.0**	**40.0**	**40.0**	**40.0**
CWP Number 02-1 - Model 1	**Designing**	Designer 3	320		40	40	40	40	40	40	40	40																		
CWP Number 02-1 - Model 2	**Designing**	Designer 3	320										40	40	40	40	40	40	40	40										
CWP Number 02-2 - Model 1	**Designing**	Designer 4	320						40	40	40	40	40	40	40	40														
CWP Number 02-2 - Model 2	**Designing**	Designer 4	320														40	40	40	40	40	40	40	40						
Sub-Total Areas 02 Designing Hours			**1280**		**40.0**	**40.0**	**40.0**	**40.0**	**80.0**	**80.0**	**80.0**	**80.0**	**80.0**	**80.0**	**80.0**	**80.0**	**80.0**	**80.0**	**80.0**	**80.0**	**40.0**	**40.0**	**40.0**	**40.0**	**0.0**	**0.0**	**0.0**	**0.0**	**0.0**	**0.0**
CWP Number 02-1 - Models 1 and 2	**Checking**	Checker 2	160																		40	40	40	40						
CWP Number 02-2 - Models 1 and 2	**Checking**	Checker 2	160																						40	40	40	40		
Sub-Total Area 02 Checking Hours			**320**		**0.0**	**0.0**	**0.0**	**0.0**	**0.0**	**0.0**	**0.0**	**0.0**	**0.0**	**0.0**	**0.0**	**0.0**	**0.0**	**0.0**	**0.0**	**0.0**	**40.0**	**40.0**	**40.0**	**40.0**	**40.0**	**40.0**	**40.0**	**40.0**	**0.0**	**0.0**
Total Area 02 Hours/Manweeks			**2640**	**66**	**2.0**	**2.0**	**2.0**	**2.0**	**3.0**	**3.0**	**3.0**	**3.0**	**3.0**	**3.0**	**3.0**	**3.0**	**3.0**	**3.0**	**3.0**	**3.0**	**3.0**	**3.0**	**3.0**	**3.0**	**2.0**	**2.0**	**2.0**	**2.0**	**1.0**	**1.0**
PLANT AREA 03																														
	Area Lead		1040		40	40	40	40	40	40	40	40	40	40	40	40	40	40	40	40	40	40	40	40	40	40	40	40	40	40
Sub-Total Area 03 Lead Hours			**1040**		**40.0**	**40.0**	**40.0**	**40.0**	**40.0**	**40.0**	**40.0**	**40.0**	**40.0**	**40.0**	**40.0**	**40.0**	**40.0**	**40.0**	**40.0**	**40.0**	**40.0**	**40.0**	**40.0**	**40.0**	**40.0**	**40.0**	**40.0**	**40.0**	**40.0**	**40.0**
CWP Number 03-1 - Model 1	**Designing**	Designer 5	320				40	40	40	40	40	40	40	40																
CWP Number 03-1 - Model 2	**Designing**	Designer 5	320												40	40	40	40	40	40	40	40								
CWP Number 03-2 - Model 1	**Designing**	Designer 6	320								40	40	40	40	40	40	40	40												
CWP Number 03-2 - Model 2	**Designing**	Designer 6	320																40	40	40	40	40	40	40	40				
Sub-Total Area 03 Designing Hours			**1280**		**0.0**	**0.0**	**40.0**	**40.0**	**40.0**	**40.0**	**80.0**	**80.0**	**80.0**	**80.0**	**80.0**	**80.0**	**80.0**	**80.0**	**80.0**	**80.0**	**80.0**	**80.0**	**40.0**	**40.0**	**40.0**	**40.0**	**0.0**	**0.0**	**0.0**	**0.0**
CWP Number 03-1 - Models 1 and 2	**Checking**	Checker 3	160																				40	40	40	40				
CWP Number 03-2 - Models 1 and 2	**Checking**	Checker 3	160																								40	40	40	40
Sub-Total Area 03 Checking Hours			**320**		**0.0**	**0.0**	**0.0**	**0.0**	**0.0**	**0.0**	**0.0**	**0.0**	**0.0**	**0.0**	**0.0**	**0.0**	**0.0**	**0.0**	**0.0**	**0.0**	**0.0**	**0.0**	**40.0**	**40.0**	**40.0**	**40.0**	**40.0**	**40.0**	**40.0**	**40.0**
Total Area 03 Hours/Manweeks			**2640**	**66**	**1.0**	**1.0**	**2.0**	**2.0**	**2.0**	**2.0**	**3.0**	**3.0**	**3.0**	**3.0**	**3.0**	**3.0**	**3.0**	**3.0**	**3.0**	**3.0**	**3.0**	**3.0**	**3.0**	**3.0**	**3.0**	**3.0**	**2.0**	**2.0**	**2.0**	**2.0**
Total All Hours/Manweeks			**11520**	**288**	4-Jan	11-Jan	18-Jan	25-Jan	1-Feb	8-Feb	15-Feb	22-Feb	1-Mar	8-Mar	15-Mar	22-Mar	29-Mar	5-Apr	12-Apr	19-Apr	26-Apr	3-May	10-May	17-May	24-May	31-May	7-Jun	14-Jun	21-Jun	28-Jun
		TOTAL DESIGNER HRS PER WEEK			**80.0**	**80.0**	**120.0**	**120.0**	**200.0**	**200.0**	**240.0**	**240.0**	**240.0**	**240.0**	**240.0**	**240.0**	**240.0**	**240.0**	**240.0**	**240.0**	**160.0**	**160.0**	**120.0**	**120.0**	**40.0**	**40.0**	**0.0**	**0.0**	**0.0**	**0.0**
		TOTAL CHECKER HRS PER WEEK			**0.0**	**0.0**	**0.0**	**0.0**	**0.0**	**0.0**	**0.0**	**0.0**	**0.0**	**0.0**	**0.0**	**0.0**	**0.0**	**0.0**	**0.0**	**0.0**	**80.0**	**80.0**	**120.0**	**120.0**	**120.0**	**120.0**	**120.0**	**120.0**	**40.0**	**40.0**
		TOTAL MANHOURS PER WEEK			**272**	**272**	**328**	**328**	**440**	**440**	**496**	**496**	**496**	**496**	**536**	**536**	**536**	**536**	**536**	**536**	**536**	**536**	**536**	**536**	**424**	**424**	**368**	**368**	**256**	**256**
		TOTAL MANPOWER PER WEEK			**6.8**	**6.8**	**8.2**	**8.2**	**11.0**	**11.0**	**12.4**	**12.4**	**12.4**	**12.4**	**13.4**	**13.4**	**13.4**	**13.4**	**13.4**	**13.4**	**13.4**	**13.4**	**13.4**	**13.4**	**10.6**	**10.6**	**9.2**	**9.2**	**6.4**	**6.4**

Figure 2–22 *Manhour Spreadsheet Example.*

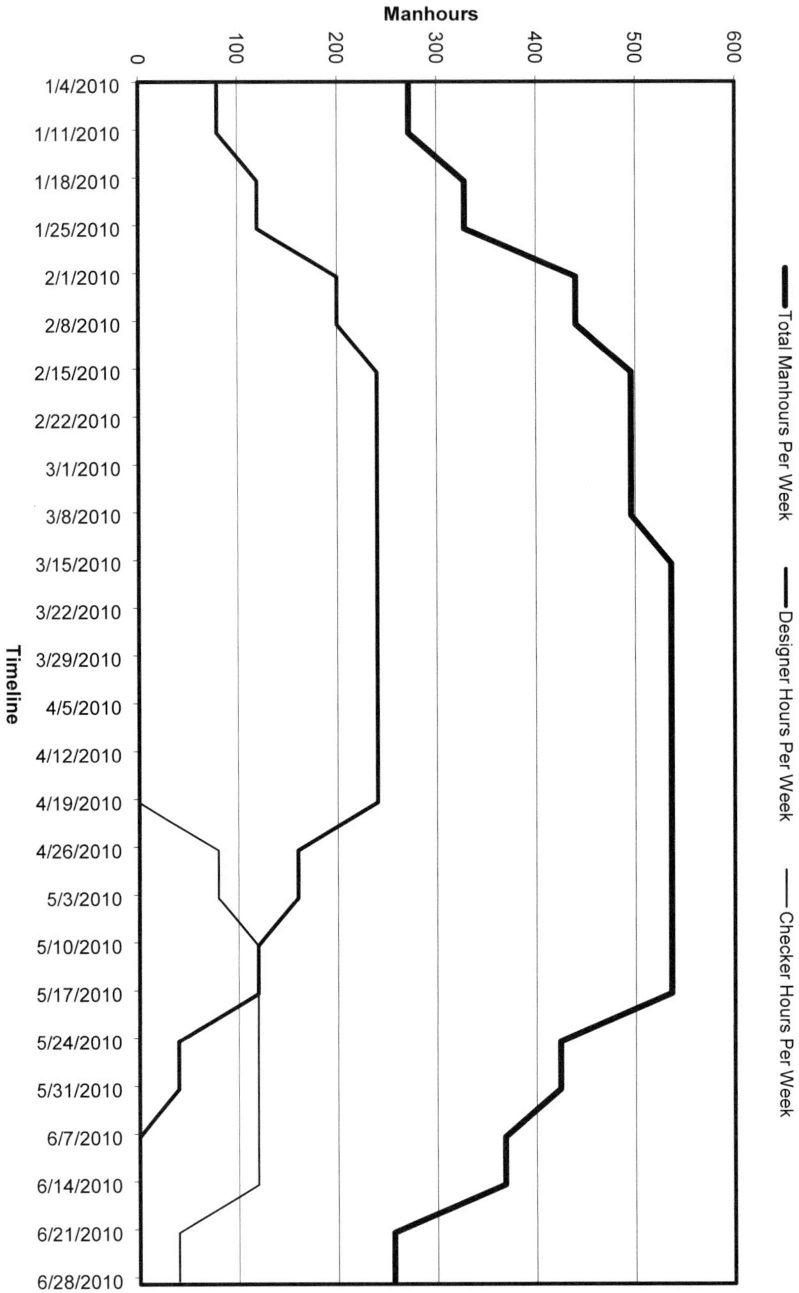

Figure 2–23 *Manhour Graph Example.*

2.8.2.4 Staffing

- Identify the area leads. Area leads will be from the ranks of the most senior designers, preferably those that have had previous lead experience.
- Now that the workload has been spread out over the duration of the project, you can clearly identify the number of designers and checkers you will need, and when you will need them. Pre-plan and assign all of the model/design areas to individual designers and checkers. If you know who the team members are, name the individuals and their assigned areas so that they know what they will be working on next.
- Take into account the level of experience you will need when assigning the work. For example, you cannot give a reactor area or distillation columns to a junior designer. More appropriately you would assign piperacks or a storage tank to the more junior personnel.
- Make sure that the plant area leads have balanced teams in terms of senior, intermediate, and junior designers. A good ratio is that for every two seniors, there will be two intermediates and one junior.

2.8.3 Other Considerations

- Do not staff up too soon. You have planned your manpower requirements to meet the schedule, but do not forget that this is contingent on receiving the information you need on time, i.e., P&IDs and preliminary vendor data. Until this becomes available, you will just be spinning your wheels and burning hours with little productivity. Start an action item list of the things that are holding you up.
- Trend all changes and rework, even though they may not be approved, and keep a record.
- Revisit and reforecast your plan a couple of times during the project.

Do not expect that your estimating and planning will go exactly as planned. They never will. They are tools intended to give you the heads-up when you're going off of the rails and give you time to react.

CHAPTER 3

Deliverables

3.1 Introduction

In this chapter we will discuss each of the possible piping deliverables and the reasons why they are produced. When we talk of deliverables from the piping group, we are of course primarily referring to the piping drawings, but I also want you to consider piping scope during this discussion.

Over the decades, industry has developed various deliverables to communicate specific aspects of the design to the other disciplines and to construction. Each deliverable brings its own added value to the project by providing the needed details to construct that aspect of the scope. The deliverables cover all the information required to build the plant, from the larger overview down to the smallest components.

This chapter will not only describe the deliverables, but will also discuss the implications of leaving any of them out. Leaving out a deliverable from the project scope to save hours, and filling in the gaps created by the omission, will usually result in more hours being wasted than if the deliverable had been included. Keep in mind that some deliverables, e.g., P&IDs, cannot be left out due to being a regulatory requirement.

All of the possible deliverables are not required for every project; this is because all of the activities that they are designed to communicate are not within the project scope. Which deliverables are required should be a simple question of "Which of the list of possible deliverables do we need to communicate the scope of our design to construction?"

Unfortunately in the real world this is not usually the question. Often the best interests of the project as a whole are overlooked and the question becomes, "How can we cut hours from the engineering budget?" This is answered with poor reasoning and/or hastily reached decisions to cut certain deliverables from the project scope, but not necessarily the scope of work itself.

Anyone can propose to address this question by offering to delete any deliverable, such as the isometrics. The key, if you are to delete a deliverable, is to understand the implications of, and weigh before hand, the pros and cons of omitting that particular deliverable. And, how you will fill the gap created by its omission.

Therefore, the question must be intelligently answered after due consideration that recognizes the value of each deliverable. As the piping lead, and a subject matter expert, it behoves you to enter into the discussions to raise flags and assist others to reach the right conclusion.

It is crucial for the project and the piping group in particular that you begin with a clear picture of your deliverables and scope. Do not expect that everyone is on the same page as you. Many projects have been started before a consensus has been reached. This is bad because your planning will be based on assumptions that may prove to be incorrect resulting in unbudgeted hours to put right later. Prior to starting your project you must engage all of the stakeholders and agree which deliverables are required and why. The major stakeholders are:

- Project management, both engineering company and client.
- Construction management.
- Piping engineering.
- Piping design.

Let's consider three scenarios of decisions made about piping deliverables and scope, and discuss the implications.

I have witnessed small projects, such as a gas pipeline dehydration installation, where the equipment is primarily vendor supplied skidded units, and there is little, other than interconnecting piping, to be designed. The decision was made to provide a plot plan, the P&IDs and Line Designation Tables (LDTs) only, and run the piping in the field, the theory being that it is simple piping and the construction could get an earlier start by cutting the engineering scope. There are some projects that have been pulled off in this manner, but to deduce that it is simple piping not requiring piping arrangements and isometrics, and that time can be saved by omitting these, may not be true. It could in fact be disastrous. When piping arrangements and isometrics are omitted as a deliverable the field personnel will have to design the routing of the piping. We have transferred design responsibility to the field, and in so doing added scope to the field effort. Before a decision is made to omit piping arrangements and isometrics consider the following:

- Is this the most efficient use of the field personnel's time? Do not forget that someone has to do the design, Material Take-Off (MTO), and material purchase before the pipe fitters and welders can get started.
- By saving on the office engineering hours, have we merely in fact increased the more costly field hours? That is, are we going to save money or will we just increase the total installed cost (TIC)? Most likely the construction schedule and costs will be increased. Time spent in the office is always a fraction of the time and cost that will be spent in the field.
- Are the skill sets available in the field to follow the P&IDs and run the piping? For example, will the field personnel recognize 'no pockets' requirements?
- How are the engineering aspects to be addressed? And, what will be the permanent stamped record? For example, stress analysis.
- The decision may entail that a full time field engineer be assigned to oversee the installation and testing.
- Most importantly, how do we ensure that the installation will be safe? Engineering companies make their money by selling engineering capabilities. Construction companies make their money by selling erection capabilities. It is not that the two do not overlap or understand each other, but it is that the agendas are different.

The theoretical and the practical invariably depart also when the field start running into problems and require support from the engineering staff. This interrupts the other projects that the engineering staff has moved on to and leads to unbudgeted hours being expended.

In this scenario, I believe it is best to have the engineers and designers do what they do best and have the field construction team do what they do best. It is not efficient to mix the two. Instead, I would argue for full engineering in the office and the omission of the piping arrangements and construction isometrics in favor of full system isometrics (on D size or A1 size sheets) and allowing the construction team to decide on the best fabrication and erection strategy.

This second scenario is not so much one of omission of deliverables, as it is one of deletion of scope from the piping group by the engineering company's project team. In the above example I mentioned vendor supplied skidded units. These are commonly almost off-the-shelf items such as in-line heaters and compressor packages that the manufacturers do on a daily basis. No one would suggest that the engineering company could or should compete and design these skidded units and the associated piping themselves. However, there are misconceptions on modularized projects that these skid manufacturers and/or module fabricators are best suited to design one-off equipment modules, and that engineering hours can be saved by placing the design in their hands. If you do not recognize and discuss the downside of this thinking, all of the design hours may well be taken out of your budget for that scope. You will need hours above the normal vendor squad checking hours to interface your designs with the manufacturer's designs and to resolve issues that will arise due to the following reasons:

- A skidded unit, if not already predesigned, is at least available as a preliminary layout at an early enough stage for the engineering company's designer to position it into the overall design to accommodate all around access. An equipment module however is not a predesigned off the shelf unit, and when the scope of a module design is removed from the surrounding plant design, effectively there are two designers working in the same area at the same time. When two designers are designing around each other there will be communication and efficiency issues to be overcome, especially when they are isolated from each other. A designer who is laying out a module with no knowledge of the surrounding area will likely block access and clearance.
- The engineering company has no control over the level of experience of the manufacturer's designers. A designer who primarily lays out skids, or primarily draws spool details for a living, may lack the experience to design more complex plant piping, and the fact that they are working in isolation from the engineering team means that communication is primarily through the vendor's drawing submissions and model reviews. When the engineering company's designers see the manufacturer's designs they will undoubtedly want to change them. Their mark-ups will not only take time, but will often become a threat of schedule delay on the part of the manufacturer. The connecting piping, having been

designed to the mark-ups may eventually be redesigned to accommodate the vendor in the interest of not delaying the module delivery.

- There will likely be retrofits required in the field to rectify access and clearance issues.
- You may have stress interface issues. Skid manufacturers and module fabricators do not usually have their own stress analysis capability and the engineering company may even have to provide the stress analysis.
- You may have heat tracing design issues. Even when the tracing design is left within the scope of one company the coordination required will be very challenging. Should it be divided between companies, the coordination required can be considerable.

Just to clarify, there are companies that provide full engineering and module fabrication capabilities. I am not talking about these companies.

Ultimately it is your decision, but I believe it is preferable to not relinquish the design of an equipment module to a skid or module manufacturer.

My third scenario is quite a simple one. Many people believe that spooling fabricators create isometrics, so why spend the time and money in the engineering office to create something that the spooling fabricator will do anyway? There is a big difference between a spool drawing or cut sheet and a construction isometric. The only similarity is that spool drawings are often also drawn in an isometric format. This will be discussed further in this chapter and in Chapter 5.

In the above scenarios I have made some observations and drawn conclusions. It is not my intent that you accept these scenarios for your own experience. My point is that the piping deliverables are not just drawings. They are inextricably linked to, and a part of, the overall project execution strategy. Taking any of them out of the equation may not save the intended design and engineering hours, and can even have an adverse effect on the project outcome. You must understand your deliverables and scope and provide input into what they will be.

Having addressed some of the possible pit-falls of leaving out a deliverable and scope, we will now move on to the deliverables themselves.

3.2 Deliverables

Piping deliverables can include all of the following:

- Cover sheets and drawing indexes.
- Plot plan.
- Key plans:
  - Piping arrangement key plan.
  - Model key plan.
  - Module key plan.
  - Construction Work Package (CWP) key plan
- Location plans:
  - Equipment location plans.
  - Tie-in location plan.
  - Utility station location plan.
  - Safety shower and eye wash location plan.
  - Heat tracing manifold location plan.
- Piping arrangements.
- Isometrics.
- Isometric logs.
- Tie-in isometrics.
- Tie-in list.
- Demolition drawings.
- Heat tracing circuit layouts.
- Heat tracing logs.
- 3-D models.
- Model indexes.
- CWP Drawing Packages and Scopes of Work (SOW).

Other deliverables worked on in many companies by the piping group are:

- Process Flow Diagrams (PFD).
- Piping and Instrumentation Diagrams (P&IDs).
- Line Designation Tables (LDTs).

While these are not strictly a piping deliverable, so much as they are a process group deliverable, the pipers have a large role to play in the drafting and management of these documents, and in particular the P&IDs.

3.2.1 Cover Sheets and Drawing Indexes

3.2.1.1 A Cover Sheet Identifies

- The project.
- The area of work.
- The CWP number.
- A description of the package.
- The reason for issuing, e.g., Issue For Bid, Issue For Construction.

3.2.1.2 Drawing Indexes List

- The CWP number.
- The types of drawing, e.g., piping arrangements and isometrics.
- The drawing numbers.
- The drawing titles.
- The revision numbers.

Each CWP drawing package will require a drawing index of the above format. There should also be a "REMARKS" column. This is used to note certain drawing anomalies, e.g., "DRAWING ON HOLD," "DRAWING REVISED," "DRAWING ADDED," and "FOR INFORMATION ONLY." The index itself should have a revision number that is incremented each time a revision to the package is issued. The drawings are to be assembled in the same order as they are listed.

Cover sheets and drawing indexes are required to help maintain the integrity of a CWP drawing package. Constructors need a list to check against, in order to verify that they have all the drawings required for their scope. The identification of a package and the drawings contained within also helps tremendously when drawings are re-issued. They help identify to whom these revised drawings are to be sent and which package they belong in. The individual CWP drawing package indexes are also used to create a master drawing list at close-out.

It must be noted that a drawing package is a dynamic (ever changing) collection of drawings, i.e., new revisions are inevitable. The drawing index is a very important document that captures these changes for the end user.

There may be more than one drawing type within a drawing package, i.e., piping packages will have piping arrangements and isometrics. In these cases list the groups of drawings separately with a heading for each in the drawing index in the same order as they are arranged within the package, i.e., "PIPING ARRANGEMENTS," "ISOMETRICS." Figure 3–1 is an example of a Drawing Index.

3.2.2 Plot Plan

A plot plan shows an overall view of the plant and the area of real estate the facility is to be built on. It identifies plant north and true north and includes a plant grid coordinate system for the location of all of the buildings, equipment, piperacks, and roads and includes an equipment list. This drawing is to scale, e.g., 1:1000. Civil and electrical will overlay their work on the plot plan to produce the foundation location plans and area classifications. Certain information is mandatory for the development of the plot plan:

- P&IDs.
- Preliminary equipment sizes.
- The plot limits.
- The lay of the land.
- Road and power access.
- Prevailing wind direction.

The development of the plot plan is based on:

- Spacing guidelines.
- Construction execution.
- Maintenance ingress and egress.
- Product shipping, e.g., rail, road, pipeline.
- Process requirements.

Plot plans are fine-tuned during the piping study stage. Figure 3–2 is an example of a plot plan.

CWP NUMBER: X
PROJECT NUMBER: X
ISSUE DATE YYYY/MM/DD
DRAWING INDEX REV. X

COMPANY LOGO

FACILITY, PAD or PIPELINE DESCRIPTION - PROJECT DESCRIPTION
PLANT AREA DESCRIPTION
PACKAGE DESCRIPTION
DISCIPLINE - DRAWING DESCRIPTION (E.g. PIPING - PIPING ARRANGEMENTS, ISOMETRICS)

NO.	DRAWING NO.	TITLE	REVISION	REMARKS
1				
2				
3				
4				
5				
6				
7				
8				
9				
10				
11				
12				
13				
14				
15				
16				
17				
18				
19				
20				
21				
22				
23				
24				
25				
26				
27				
28				
29				
30				
31				

Figure 3–1 *Drawing Index.*

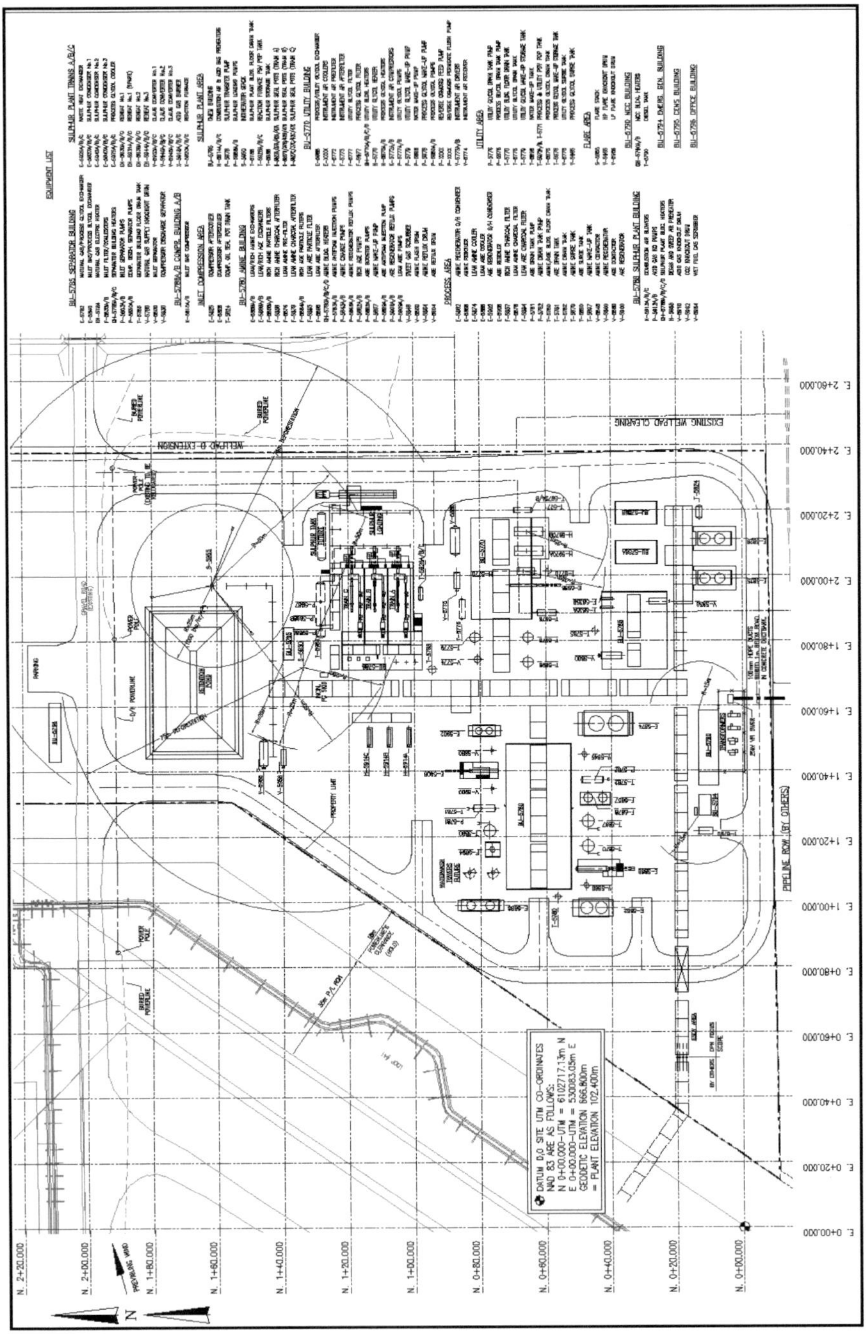

Figure 3–2 *Plot Plan.*

3.2.3 Key Plans:

All key plans have several things in common:

- They are an overlay to the plot plan.
- They are important reference documents for relating the subject matter boundaries.

Figure 3–3 is an example of a key plan, specifically a piping arrangement key plan.

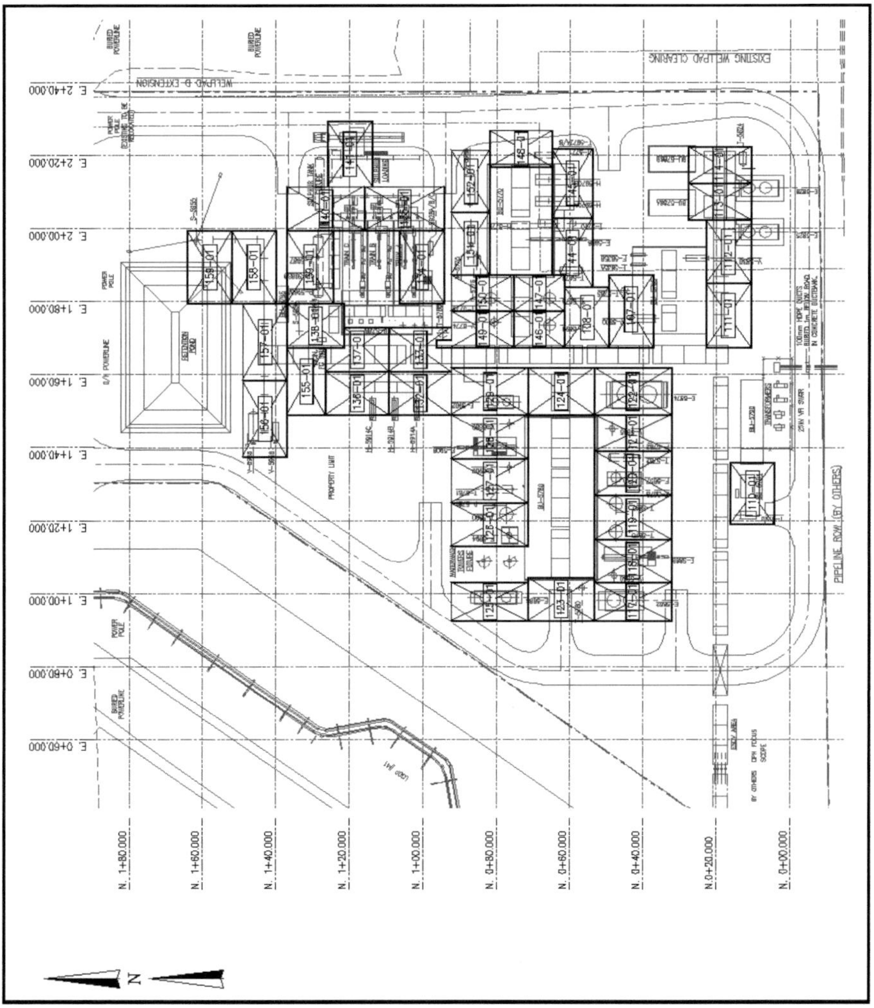

Figure 3–3 *Piping Arrangement Key Plan.*

3.2.3.1 Piping Arrangement Key Plan

The piping arrangement key plan shows the matchline limits and the drawing numbers of each of the piping plans that will be drawn at a 1:30 scale on an ANSI D or ISO A1 size sheet of paper.

3.2.3.2 Model Key Plan

A piping model key plan shows the limits of each of the models and the model numbers. All disciplines will create their own model key plan, and the primary use during detailed design is as a reference for the designers to know which models they are required to externally reference into their own models. Later they become an important document during construction for locating the models of a given area.

3.2.3.3 Module Key Plan

It is important to have a module key plan. A module key plan not only references the limits of the modules and the module numbers, it is also used for construction planning, such as the movement and lay-down of modules and critical lifts.

3.2.3.4 Construction Work Package Key Plan

The construction work package key plan shows the limits of the CWPs and the CWP numbers. As with the module key plan, it is an important document for construction planning.

3.2.4 Location Plans

All location plans are created for quick reference as to the location of the subject matter, and, with the exception of the equipment location plans, they are an overlay to the plot plan.

3.2.4.1 Equipment Location Plans

Plot plans of larger facilities are often at too small a scale for all of the smaller pieces of equipment to be located by co-ordinates and equipment tag numbers due to the physical constraints of placing the text on the drawing. Therefore, equipment location plans are created at a larger scale, e.g., 1:400. Several equipment location plans are required to cover the same area as the plot plan, each with a list of the equipment located within the drawing. The list includes:

- Equipment tag number.
- Equipment description.
- Coordinates.
- Elevation.

Equipment location plans simplify the locating of the individual pieces of equipment.

3.2.4.2 Tie-In Location Plan

As a companion to the tie-in list, the tie-in location plan will show the tie-in numbers. Commonly the tie-in numbers are noted within a hexagon and a leader line from the hexagon points to the approximate plant location. Because a typical project with tie-ins may have dozens of tie-ins, this is a handy reference for identification and planning.

3.2.4.3 Utility Station Location Plan

A utility station location plan indicates the utility stations by tag number at the approximate plant locations. Often a symbol for the type of utility station and a legend are employed within the drawing, along with a list by tag number to identify each of the utility stations:

- Type, i.e., the combination of utilities servicing the station: air, nitrogen, steam and water.
- Coordinates and elevation.

3.2.4.4 Safety Shower and Eye Wash Location Plan

A safety shower and eye wash location plan indicate the safety shower or eye wash stations by tag number at the approximate plant locations. Similar to the utility station location plan, a symbol for the type of safety shower/eye wash and a legend are often employed within the drawing along with a list by tag number to identify each of the safety shower and eye wash stations:

- Type, i.e., safety shower, eye wash, or both.
- Coordinates and elevation.

This is a document related to safety, and you will often see them pinned to the walls in the operations room and around the plant.

3.2.4.5 Heat Tracing Manifold Location Plan

A heat tracing manifold location plan indicates the manifolds by tag number at the approximate plant locations. Also, similarly to the utility station location, and eye wash and safety shower location plans, a symbol for the type of manifold and a legend are employed within the drawing along with a list by tag number to identify each of the manifold stations:

- Type, i.e., horizontal or vertical, number of connections, supply or return.
- Coordinates and elevation.

3.2.5 Piping Arrangements

Piping arrangements are also known as piping layouts. They are drawn at a 1:30 scale and show all equipment, pipe routings, valves, instruments, shoes, anchors, guides and attached supports, and supporting steel locations. Dimensions, elevations and coordinates are included. Other information also shown on these drawings are the instrumentation and the platforms and ladders. The purpose of the piping arrangements is that they enable the user to visualize the overall relationship of all the piping systems, steel and equipment. The lack of piping arrangement drawings will force construction personnel to create their own layout sketches by reference to multiple isometrics. While screenshots of the 3-D models will help, it can be a time consuming exercise to recreate the same information in another format.

There are two types of piping arrangements:

- Area piping arrangements
- Module piping arrangements

Figures 3–4 and 3–5 are examples of piping arrangements.

3.2.6 Isometrics

Isometrics are referred to as isos for short. These drawings are produced so that the piping can be broken down into manageable line segments. They are not drawn to scale, but will be of reasonable proportion. Because an isometric shows the routing of one line only, in a 3-D format, they are easy to read. They are also smaller drawings (ANSI B or ISO A3) making them easy to handle. A "Bill of Material" (BOM) is included on each isometric for the fabrication and erection

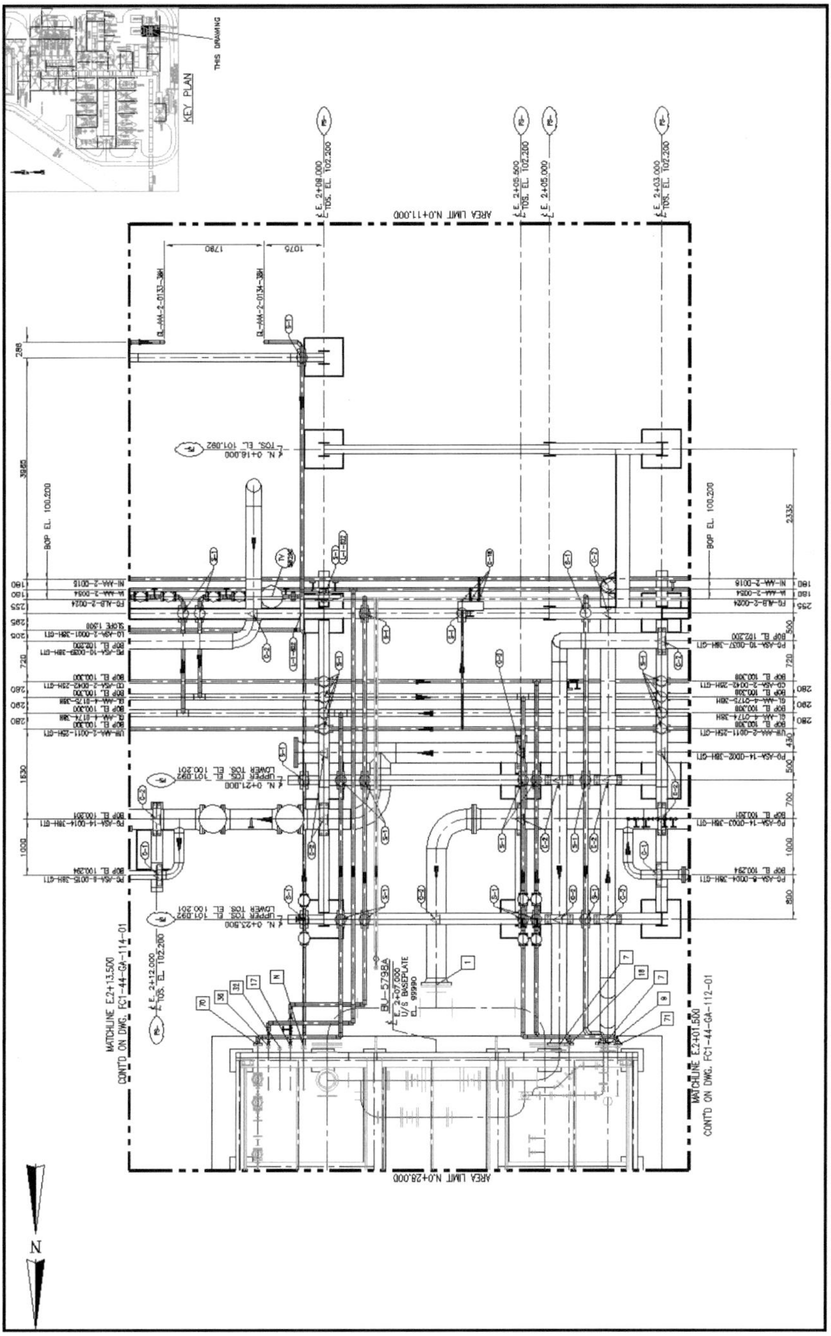

Figure 3–4 *Area Piping Arrangement.*

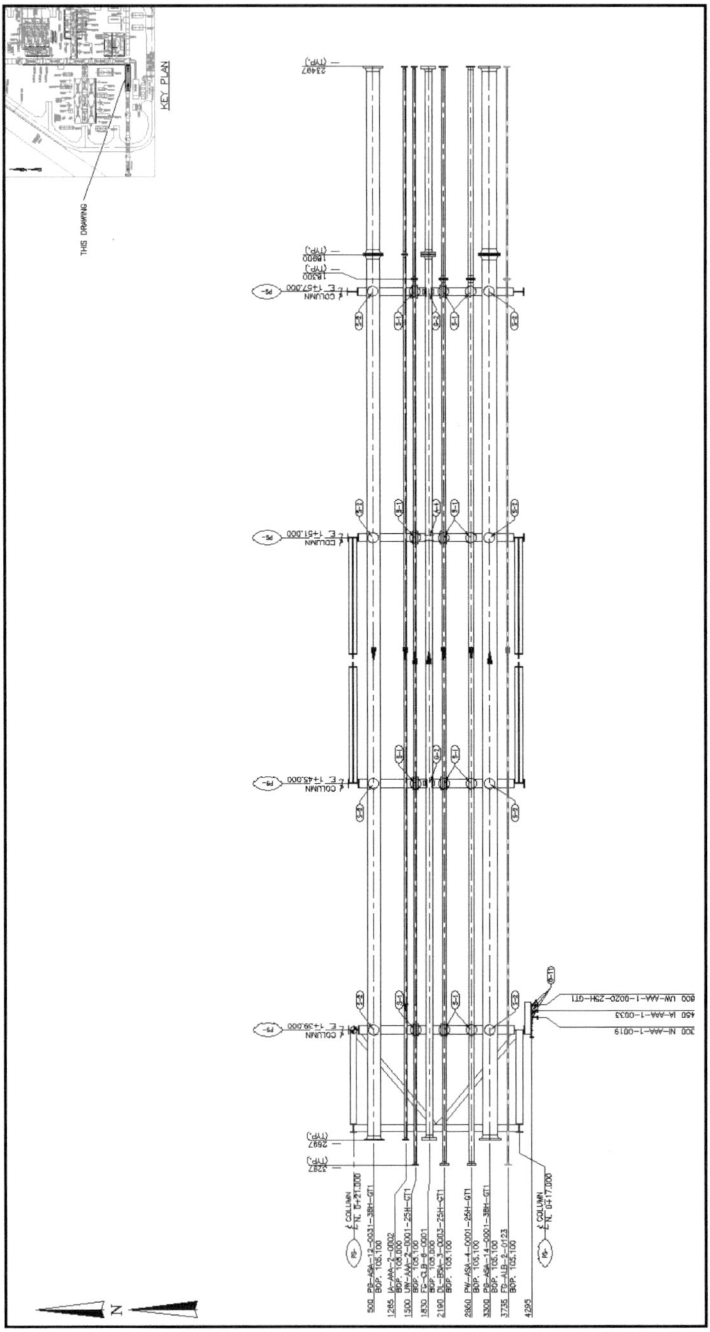

Figure 3–5 *Module Piping Arrangement.*

of the line. The BOM identifies whether delivery of material is to be directed to the shop or to the field. They also include design, operating and test conditions, taken from the LDT.

The spooling fabricators use them for estimating and planning the work and follow the isometrics to produce their own spool sheets. You should be aware that fabricators will charge less to produce spool sheets when isos are provided than they will charge when they have to work from piping arrangements only.

The electrical group use isos as background for electrical heat tracing drawings. The field construction use isos for:

- Identifying and assembling the spool pieces.
- Progress monitoring.
- Weld mapping.
- Setting up hydrotest packages.
- Walk down of the lines and identification of deficiencies (punch lists).
- Testing and commissioning.

The isometrics are another important tool to be provided for fabrication and construction from the engineering company. Without isometrics, fabrication and construction will be obliged to create their own sketches to divide the work down into the components of single lines for BOM take-off, work allocation, and work execution, and will have to search elsewhere for installation and testing requirements. Figure 3–6 is an example of an Isometric.

3.2.7 Isometric Logs

Isometric logs are used to list and track all of the isometrics. The two logs you require cross reference each other and are:

- One that lists all of the isometrics in numerical order and the line numbers that are included within each isometric.
- One that lists all of the line numbers and the isometric that the line may be found on.

3.2.8 Tie-In Isometrics

Prior to your regular construction isometrics being issued, it will be required to issue the tie-in isometrics. There are no real differences between a tie-in isometric and any other isometrics. The exception is

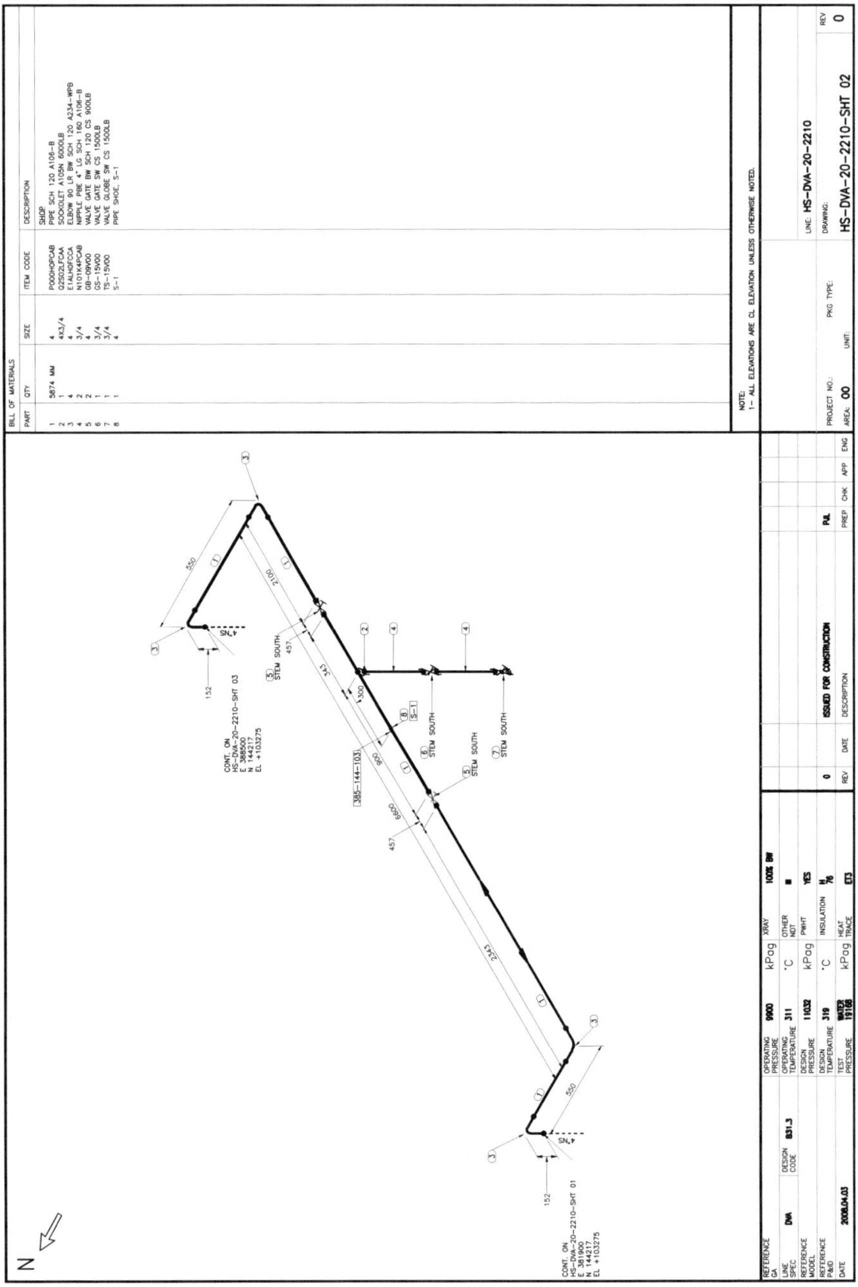

Figure 3–6 *Isometric.*

that they detail only the tie-in and temporary material requirements that will be removed once the continuation piping is connected. For exmaple, a blind flange and a bleed connection are often added to the isolation valve for testing the soundness of the valve prior to the continuation being connected.

3.2.9 Tie-In List

Tie-in lists identify the tie-in numbers assigned to the locations where a new piping system is to be joined to an existing piping system. The tie-in numbers are shown on the P&IDs, the piping arrangements, and the isometrics. The tie-in list includes:

- The tie-in number.
- The reference drawings, i.e., P&ID, piping arrangement, isometric.
- The type of tie-in, e.g., hot tap.
- Any special engineering instructions to the field.

The piping group will assign the tie-in numbers and fill-in the reference columns. The piping engineer will populate the engineering data. Projects with tie-ins require a field trip to establish the type of tie-in with operations, tag the tie-in locations and route the continuation piping from them through the existing area. Figure 3–7 is an example of a tie-in list.

3.2.10 Demolition Drawings

Demolition drawings are required as an aid to construction for revamp projects in an existing facility where existing piping is to be removed. They are copied/created from the existing P&IDs and piping arrangements prior to any revisions being done for the new piping that will be installed. They are an aid to construction and are created in order to show the existing piping that must be removed. The piping to be removed is cross-hatched, and includes the tie-in numbers where new piping will be tied in. The tie-in numbers shown correspond with those found on the P&IDs, tie-in list, piping arrangements and isometrics.

While they are a deliverable for the project, they are not usually a deliverable at close-out. At close-out the client is only interested in receiving the revised as-built drawings. Figure 3–8 is an example of a demolition drawing.

TIE-IN NUMBER	LINE, EQUIPMENT, INSTR. EXISTING	LINE, EQUIPMENT, INSTR. NEW	P & ID DRAWING NUMBER	PIPING GA DRAWING NUMBER	ISOMETRIC DRAWING NUMBER	SUPPLEMENTAL ENGINEERING INSTRUCTION

Figure 3–7 *Tie-In List.*

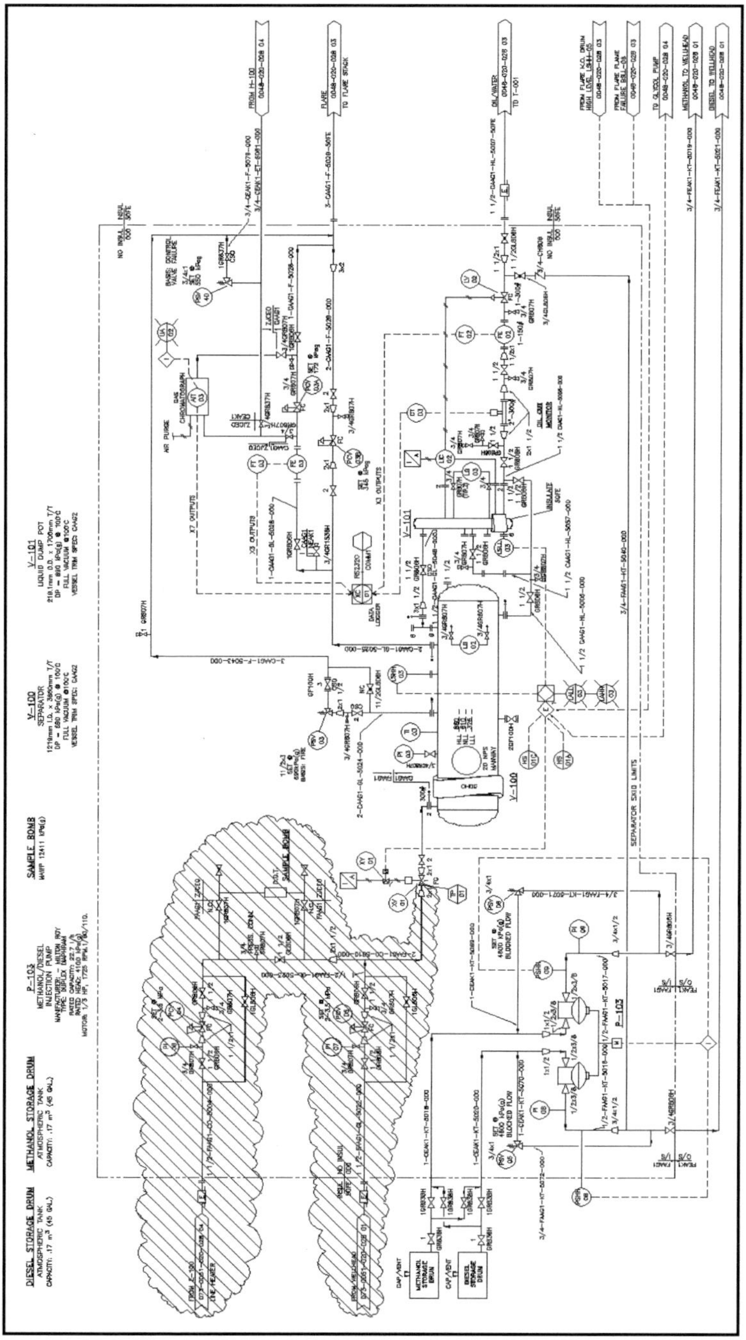

Figure 3–8 *Demolition Drawing.*

3.2.11 Heat Tracing Circuit Layouts

The heat tracing, be it by medium of hot oil, steam or glycol, is there to provide heat for freeze protection or process requirements. Electrical tracing may be used also, but it will not be a part of this discussion of piping deliverables.

For the heat tracing circuit layouts, the piping arrangements and isometrics are used as a background and the circuits are drawn using the heat trace specifications and standards. The circuits are drafted using polylines, originating from the supply manifold, along the traced line, and terminating at the return manifold. Tag numbers and flow arrows are used to identify the circuits. The manifold numbers are also indicated on the heat trace drawings.

It has been my experience that this is a task for which adequate hours are often not provided. It has also been my experience that this is a prime example where the transference of a task to the field to save engineering hours, results in increased overall project hours.

Additionally, the decision to make heat tracing design part of the field construction process often leads to designing on the fly that can result in poor circuit design: tracer lengths that are exceeded, wrong number or size of tracers, dead legs, and insufficient record keeping. Once a line has been insulated it is difficult or impossible to determine the tracing. Design in the office will save time in the field and will result in a superior product and a permanent record.

3.2.12 Heat Tracing Logs

Heat trace logs are the companion to the heat trace drawings and describe the individual tracer number, the originating and terminating manifolds, the piping line number that is traced, and the tracer length and size.

You will require three heat trace logs:

- One for each supply manifold, which lists the tracers originating from that manifold, the lines traced, and the return manifold.
- A companion log of each return manifold that lists the tracers, the lines traced, and the supply manifold.
- A cross reference log by line number that lists all of the tracers associated with that line.

Figure 3–9 is an example of a heat trace log.

Tracer Number				LINES/EQUIPMENT/ INSTRUMENT TRACED (INSERT ADDITIONAL ROWS IF MULTIPLE LINES ARE TRACED BY A SINGLE TRACER)	Return Connection Number				REMARKS
AREA	SM - SUPPLY MANIFOLD	SUPPLY MANIFOLD SEQUENTIAL NUMBER	VALVE CONNECTION NUMBER		AREA	RM - RETURN MANIFOLD	RETURN MANIFOLD SEQUENTIAL NUMBER	VALVE CONNECTION NUMBER	
0X-	SM-	XX-	01		0X-	RM-	XX-		
0X-	SM-	XX-	02		0X-	RM-	XX-		
0X-	SM-	XX-	03		0X-	RM-	XX-		
0X-	SM-	XX-	04		0X-	RM-	XX-		
0X-	SM-	XX-	05		0X-	RM-	XX-		
0X-	SM-	XX-	06		0X-	RM-	XX-		
0X-	SM-	XX-	07		0X-	RM-	XX-		
0X-	SM-	XX-	08		0X-	RM-	XX-		
0X-	SM-	XX-	09		0X-	RM-	XX-		
0X-	SM-	XX-	10		0X-	RM-	XX-		
0X-	SM-	XX-	11		0X-	RM-	XX-		
0X-	SM-	XX-	12		0X-	RM-	XX-		
0X-	SM-	XX-	13		0X-	RM-	XX-		
0X-	SM-	XX-	14		0X-	RM-	XX-		
0X-	SM-	XX-	15		0X-	RM-	XX-		
0X-	SM-	XX-	16	SPARE	0X-	RM-	XX-	SPARE	
0X-	SM-	XX-	17	SPARE	0X-	RM-	XX-	SPARE	
0X-	SM-	XX-	18	SPARE	0X-	RM-	XX-	SPARE	
0X-	SM-	XX-	19	SPARE	0X-	RM-	XX-	SPARE	
0X-	SM-	XX-	20	SPARE	0X-	RM-	XX-	SPARE	

REV NO.	DATE	DESCRIPTION	BY	CHKD	APPD

PROJECT	SUPPLY MANIFOLD TRACING SCHEDULE		
PROJECT NO.	DOCUMENT NUMBER	SHEET	REV

Figure 3–9 *Heat Trace Log.*

3.2.13 3-D Models

Of all the deliverables that can be provided, the 3-D models are often initially the least understood, the least discussed, and later become the most controversial. Too often clients assume that they have paid for the models, but if they are not in the contracts as a deliverable the engineering company may not be willing to hand them over. Even when they are listed in the contracts or the engineering company is willing to hand them over regardless, they may still be unwilling to hand over the customizations and the configurations that they have made to the software. In effect this can render the models almost unusable for future expansion work by another engineering company.

Assuming the client wants the models, which is the first thing to determine, I can only offer the following advice:

For those on the engineering company side of the fence:

- Clearly define the software you will be using, the version of that software, and what will not be included in the turnover but can be available at an additional pre-described cost. You do not want to lose the possibility of future work due to a disagreement with your client that can be anticipated and avoided ahead of time by discussion and written documentation.

For those on the client side of the fence:

- Understand that the models and databases are many faceted beasts. Merely stating that you want the models turned over at the end of the project will not be sufficient.
- Document the software, version of software, your turnover requirements, and your expectations in terms of usability after turnover.
- If you lack the in-house expertise, engage an outside independent consultant to help you define the parameters of the turnover.

3.2.14 Model Indexes

As per the drawing indexes, you must create model indexes that list all of the models generated for the project and they must concur with the model key plan. This is an invaluable document at the time of close-out for the project and is often a requirement from the client. Ensure that you have a list that is current at all times.

3.2.15 CWP Drawing Packages and Scopes of Work (SOW)

A Construction Work Package is a well defined scope of construction work that produces a specific deliverable. The deliverable is defined in enough detail that the end result is clear, as are the budget and schedule associated with the work. As a result, a properly prepared and executed Construction Work Package, or CWP, is very manageable.

The CWP may be used as the basis for a contract between the owner and the construction contractor because of its high degree of definition and manageability. The scope of work in a work package, including a CWP, should not overlap that of another work package.

Common CWPs include:

- A description of the scope of the work to be performed, accurately defining the limits. Each CWP may require several SOWs for the completion of the package.
- The work that is not included. A description of any associated work which is not included in the contractor's scope for the CWP, i.e., activities that are required to complete the CWP that will be conducted by others.
- A list of the materials and services that will be supplied. The list will identify all materials and services that will be supplied, including when, to where, how, and in what quantities.
- A statement to the effect that the contractor is responsible for all other materials and services required to complete the work.
- A schedule for the work that identifies the expected start and completion dates, together with any intermediate milestones that must be achieved.
- Quality Assurance (QA) and Quality Control (QC) requirements.
- The CWP drawing package—Issued For Construction.
- Reference drawings stamped "FOR INFORMATION ONLY."
- A list of the company standards and specifications to be used during construction.
- A list of the vendor drawings (if required) and other attachments.

3.2.15.1 CWP Drawing Packages

It is very important that consideration be given to the CWP boundaries and drawing packages that are to be included in each of the CWPs in order to provide clarity and avoid confusion for the fabricators and erectors.

For this you must plan ahead from the very moment you are dividing up the plot plan. I have two simple rules that I follow:

- Only include the construction drawings in the CWP drawing package. Do not include the project drawings. Project drawings are drawings that do not have a CWP boundary and cover all plant areas, such as P&IDs and plot plans. These are to be issued separately.
- A drawing must never be included in more than one CWP drawing package. This requires a split between module and field erected piping drawings. For instance, do not create a piping drawing that includes both a piperack module and the field continuations. Create two drawings: one with the piperack piping detailed and the field continuations greyed out for reference, and a second for the field continuations with the module piping greyed out for reference.

The following is a description of the way I have packages assembled. As is the thread throughout the book, you may well choose a different approach, but, be sure you give due consideration to this beforehand.

There are three basic piping construction drawing package arrangements:

- Module piping.
- Field erected piping.
- Heat tracing.

There are four basic categories of drawings:

- Project drawings that cover all areas.
- Construction drawings.
- Standard drawings.
- Vendor drawings.

The management of each is covered in the following paragraphs.

3.2.15.1.1 Project drawings

Project drawings are drawings that cover all areas. These are to be assembled into separate packages with a cover sheet and drawing index. The following list covers the types of drawings that fall into this category:

- PFDs.
- P&IDs.
- LDTs.
- Plot plans.
- Key plans.
- Location plans.

PFDs are to be assembled into one package with a cover sheet and drawing index.

P&IDs are to be assembled into packages by process area, one package for each with a cover sheet and drawing index.

LDTs are to be assembled into packages by process area, one package for each with a cover sheet.

Plot plans, key plans and location plans are to be assembled into one package with a cover sheet and drawing index.

Project drawings are not included in the CWP drawing packages because they cross CWP boundaries. Were they to be included, the same drawings would have to be a part of multiple CWPs, and apart from the extra paper and handling involved, the real risk exists that it is impossible to keep track of all of the packages these drawings are included in, and revisions would never be issued for all of them. It is simpler and safer to have them in a separate package.

3.2.15.1.2 Construction drawings

Construction drawings are the engineering company's stamped (IFC) drawings used for fabrication and erection. These drawings are assembled into the individual CWP drawing packages. Each CWP package of drawings is to be a stand alone set with a cover sheet and drawing index.

- Piping arrangements
- Isometrics
- Heat tracing circuitry

3.2.15.1.3 Standard drawings

Standard drawings are the typical fabrication and/or installation details of common items. Standards come from two sources: the client and the engineering company. All standards are to be found in the project set of standard drawings. Commonly, standard drawings are not to be included in the CWP packages as contractors will receive a project control copy set as part of their award documentation.

3.2.15.1.4 Vendor drawings

Vendor drawings are the detail drawings provided to the engineering company for approval by an external supplier of purchased equipment. They are used by the external supplier for the fabrication of this equipment. These take the form of such items as pumps, vessels, tanks, exchangers, specialty items, and instruments. They can also be complete skidded units such as compressors. Generally speaking, the external supplier will ship these items to the field or to a module fabrication shop to be installed by others. The installer requires copies of the vendor drawings in order to cost and plan the execution of the installation.

Vendor drawings are issued as a listed attachment to the CWP. A cover sheet and drawing index are not required.

3.2.15.2 Scopes of Work (SOW)

A part of the CWP is the Scope of Work (SOW). The SOWs are the instruction to the spooling fabricators, module assemblers, and field piping and equipment erectors as to the scope that has been contracted to them, and as such must be very explicit in their description.

Depending on the construction and contract award strategies, the same CWP may have several different SOWs. These different scopes are for the different activities required for the completion of the CWP and are identified by the Work Breakdown Structure (WBS). For instance, there may be several SOWs associated with a field erected piping CWP:

- Fabrication and supply of the piping spools.
- Field erection of the spools and other non-welded components, e.g., flanged valves.
- Field repair of painted pipe.
- Field heat tracing.
- Field insulating of the piping.

It is quite possible that as the piping lead you will be required to have input into, if not to write entirely, the piping SOWs and assemble the CWP drawing packages. As it is most likely that someone else will assemble the packages on your instruction, you should review the packages before issuing. At the same time you should review that the general layout and the content of the drawings are clear and concise in their instruction, i.e. ask yourself, "Would I be able to build from these drawings?"

(See Chapters 5 and 6 for sample SOWs.)

3.2.16 Process Flow Diagrams (PFDs)

Process Flow Diagrams (PFDs) identify the major process streams (lines), equipment, and controls. A companion chart, often included within the PFD, is the mass flow balance. PFDs are a flow diagram, without a scale and are of value to the piping group in that they provide an overall view of the process and are the basis for the P&IDs. Figure 3–10 is an example of a PFD.

3.2.17 Piping and Instrumentation Diagrams (P&IDs)

Piping and Instrumentation Diagrams (P&ID) are "the road map" as far as the piping group is concerned. They are the drawings that allow the piping design to move forward. Consequently, they can also be the bane of every piping designer if they are changing throughout the project. Changes that will appear to be small to the process group can be significant to the piping group.

P&IDs are flow diagrams that do not have a scale. They include information of great importance to the piping designers. They show all equipment, valves, instrumentation and controls logic, specialty items, line and equipment isolation requirements such as double block and bleed, spectacle blinds and car seals, and note all process requirements, e.g., piping that must not be pocketed, and process vents and drains. They also identify each line by a line number which identifies the line size, piping class, commodity, and insulation and heat tracing requirements. Other needed information, such as vessel liquid levels, vessel tangent elevations, PSV requirements, and instrument types and sizes are also included.

In every project there will come a time, usually at the IFE stage, when the P&IDs will come under the auspices of the piping group. This is because the initial process layout will not reflect the developing piping layout of headers, sub-headers, and branches. It is required that the piping group mark-up the actual sequence of the take off of branches from the headers, and identify the need for sub-headers when two or more separate lines of the same commodity branch into the same area according to the P&ID. Obviously, the

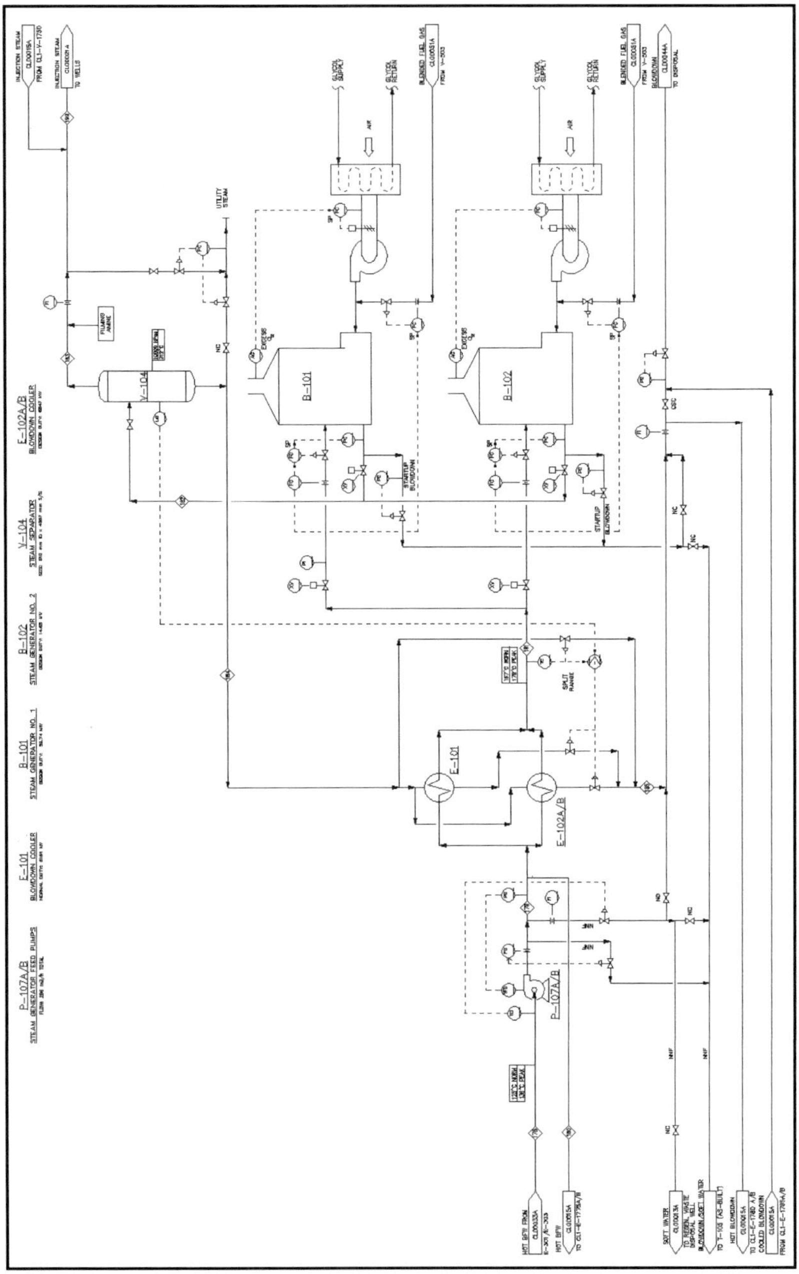

Figure 3–10 *PFD.*

piping group will not change the process, as this is clearly not their expertise, but they do need to bring the realities of the piping configurations to these drawings. Figure 3–11 is an example of a P&ID.

3.2.18 Line Designation Tables (LDTs)

The Line Designation Tables (LDTs) are created by the process group and are later maintained by the piping group at the same time as the P&IDs are turned over. The two documents go hand-in-glove and the marking up of the P&IDs will often require the marking-up of the LDT.

LDTs list the:

- Process commodity.
- Piping specification.
- Line number.
- Line size.
- Insulation requirements.
- Heat tracing requirements.
- Pipe wall schedule.
- Origin and termination of the line.
- P&ID reference (originating P&ID only).
- Operating conditions.
- Design conditions.
- Stress analysis requirements, if different from design conditions, e.g., steam out.
- Testing requirements, e.g., hydrotest and X- Ray.
- Code.

Figure 3–12 is an example of an LDT.

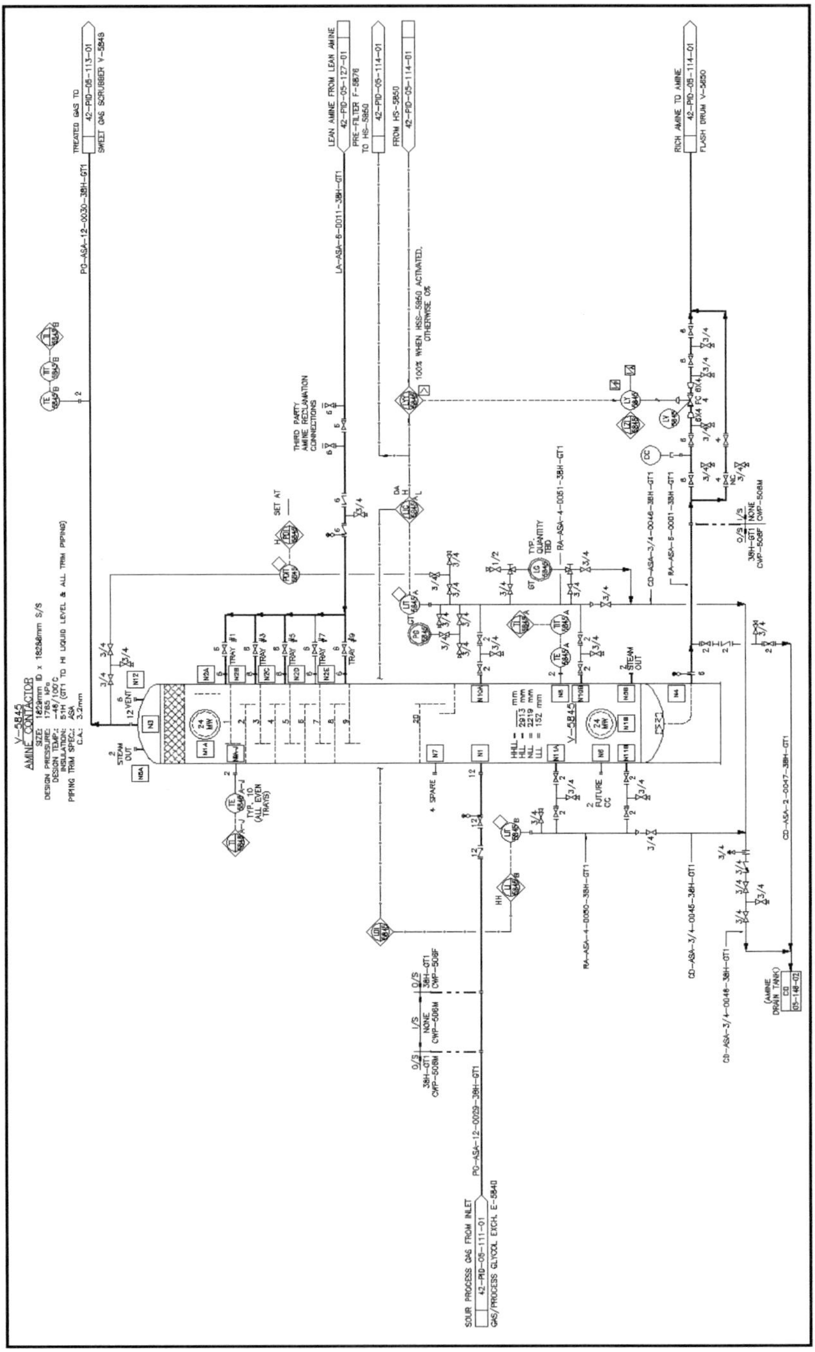

Figure 3–11 *P&ID.*

LINE DESCRIPTION					INSULATION		TRACING DESIGN			LINE ROUTING		ORIGINATING P&ID NO.	SCH OR WALL THK	FLUID PHASE	OPERATING		DESIGN		TEST		EXP TEMP	UPSET P&T FOR STRESS CALCS		MDMT TEMP	CORR ALLOW	X-RAY	NDE CAT	STRESS REL.	PAINT'G SYSTEM	INTERNAL COATING, LINER	CODE	NOTES
AREA ID	COMM CODE	CLASS SERVICE MATERIAL	SIZE NPS	LINE NO.	THK (mm)	TYPE	TYPE	HOLD TEMP °C	NUMBER TRACERS	FROM	TO		mm	V/L/M	PRESS kPag	TEMP °C	PRESS kPag	TEMP °C	PRESS kPag	MEDIUM	°C	PRESS kPag	TEMP °C	°C	mm	°C						

Figure 3–12 *LDT.*

CHAPTER 4

Detailed Design

4.1 Introduction

So far we have discussed all of the pre-planning required for you to set yourself up for success. As the piping lead you will have many issues to deal with throughout the detailed design stage of your project. This is the most intense phase of your project, where the bulk of your manhours will be expended.

All of your preparation and planning will now pay off as the interruptions that would otherwise have taken you away from the piping execution will be minimal. Having already addressed the aforementioned preparations will enable you to be more focused on the detailed design.

Recapping the previous chapters, the following have been established:

- Standards to be used.
- Specifications to be used.
- Procedures to be used.
- Software to be used.
- Manhour estimates and manpower planning.
- The deliverables.

But, there is still a lot to do in the way of planning and maintenance. The detailed design involves putting into play all of the decisions made and following through with other decisions. In this chapter, and the following two chapters, we will discuss the execution of the detailed design and highlight other activities and decisions that you will be required to make or have input into.

Because they are so important to the success of the project, and because not everything will have been firmed up entirely during the planning and estimating stage, we will begin by recapping in more detail some of the previously discussed topics that may have to be finalized during the study phase of detailed design.

4.2 Contracting and Procurement Plan and Construction Execution Plan

As has been previously discussed in Chapter 1, the Contracting and Procurement Plan and the Construction Execution Plan contain the procurement and construction philosophies and are the basis from which other decisions and planning by you, the piping lead, will stem. These are:

- Modularized and field-erected piping splits.
- Construction Work Package (CWP) boundaries.
- Model boundaries.
- Shop and field material Splits.
- Procurement splits.
- CWP drawing packages.
- Scopes of Work (SOW).

It is unlikely that the contracting, procurement, and construction plans will list all the details. For instance, it may be stated in the contracting and procurement plan that the engineering company will purchase all NPS 2 and above pipe, fittings, and valves. It may be stated in the construction execution plan that modularization will be utilized on the project. But you will have to help to decide the details of the procurement split and what can be modularized.

4.3 Modularized and Field Erected Piping Splits and CWP Boundaries

The plot plan is divided into the general process areas and each process area is broken into CWP boundaries. Equipment that falls within a general process area boundary that is tagged according to a different process will be a part of the general process area that it is located in. That is, do not create a separate CWP for a piece of equipment just because of its process designation. Commonly, a piece of equipment tagged for a different process than the area it sits in will be erected at the same time as all the other piping and equipment in that area.

The identification of the CWP boundaries should be done with the view in mind of providing clearly defined SOWs for fabrication and erection, and also for flexibility in the awarding and management of the contracts.

- An obvious boundary is the boundary between modularized and field erected piping due to the different construction philosophies. Once you have decided on these splits you can then decide on further divisions.
- Dividing a large area into several CWPs instead of having one large CWP will allow greater flexibility in the awarding and managing of contracts. With smaller packages it is possible to issue on an ongoing basis or issue to more than one fabricator.

  For instance, it may be desirable to break a long run of piperack modules into two CWPs rather than one CWP. You can choose to issue the two CWPs to one fabricator at intervals or to two different fabricators at the same time. If you have issued both the CWPs to one fabricator who is not performing, it is an easier matter to pull one CWP back and re-issue it to another fabricator than it would be to start dividing the work of the larger CWP.

- It is easier to monitor the progress and manage the schedules of smaller work packages than larger work packages.

The above is best illustrated by an example. Refer to Figure 4–1 CWP Boundaries. This example uses modularized piperacks, an equipment module and a field-erected piping area.

4.3.1 Notes to Figure 4–1

1. CWP boundaries are established according to fabrication and erection methodology. That is, modules are assembled as complete units in a shop. Field-erected piping involves spooling in a shop and later field erection.

2. CWP boundaries are not always "straight lines" that cannot be transgressed. In this example, the expansion loop is included in the pipe rack module CWP because it is needed in the field during the erection of the pipe racks.

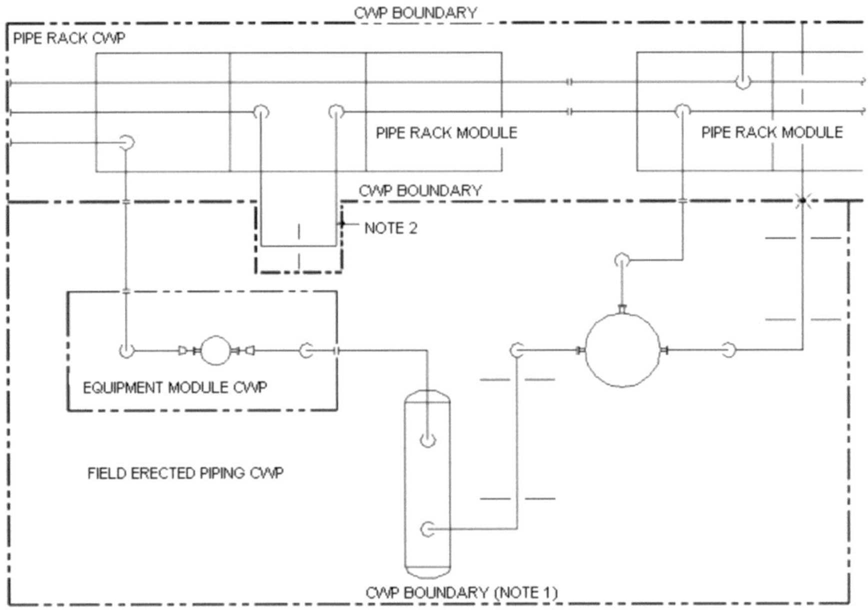

Figure 4–1 *CWP Boundaries.*

4.4 Model Boundaries

The establishing of the CWP boundaries precedes the establishing of the 3-D model boundaries. This is followed by the establishing of the piping arrangements and isometric boundaries. The model and drawing boundaries will therefore align with the CWP boundaries.

A simple rule to follow when deciding on model boundaries is to break models along the same lines as the components will be built. In other words, model it as you will build it. A prime example of this is separate models for separate modules. You must discuss this with your counterpart in the structural group as they must follow the same model boundaries as yourself.

Clearly defined boundaries equate to clearly defined information flow and good communication. The downstream team members are less likely to make misinterpretations. For instance, the Material Take-Off (MTO) reports will accurately reflect the CWP, which in turn vastly increases the chances that the material quantities are accurately purchased and shipped to the correct locations. Blurred boundaries inevitably lead to confusion and added costs.

The breaking down of a CWP into several models also allows segregation of work between designers. Designers can be assigned to each model area and the designing may be executed simultaneously, with the work having been divided sufficiently to meet the required IFC dates in the timeframe allotted.

General Instruction:

- The boundary of a piping model will be determined by:
  - One CWP boundary. Smaller CWPs will not be combined to form one larger model.
  - Module boundary. Each module must be in a separate model.
  - The schedule and manpower plan. Separate models are required for areas that will be issued in stages, and for CWP areas where the schedule dictates that two or more designers must be working simultaneously.
  - Size of the model. Models that are too large may have a tendency to crash.
- Each piping model will require an equipment model separate from the piping model. Vendor supplied piping on skids must be modeled in the equipment model so that this piping does not show up in the MTO report.

I strongly believe in a study phase of P&ID transpositions and study models prior to the creation of the detailed design models. Designing must be done in stages that build upon each other.

In the manual drafting days it was common practice to start designing by the creation of a transposition. This is the routing of the critical lines on the plot plan, done in single line format without scale and without detail. The line numbers are indicated, and control valves and flow instruments are identified by an instrument bubble. The purpose of this is to:

- Establish the basic piping arrangement. The piping within the racks, sub-racks and directly between equipment.
- Establish a preliminary piperack line sequence, rack spacing and number of levels.
- To firm-up the equipment locations. Long runs of the expensive piping, i.e., high temperature, high pressure, and alloys may be shortened by moving equipment.

I witnessed a 3-D project where a transposition was not conducted. At a congested interface section of the main rack between two sub-racks, the number of lines started to crowd out of the main rack onto outriggers. Eventually it was realized that the outriggers were not going to work and a new section of rack was added beside the main piperack, and the whole section had to be re-designed. This was not good midway through the project, especially when this could have been identified as a problematic area at a much earlier stage.

The study stage follows the transposition. Referring to the manual drafting days again, the designers would begin their layouts on vellum paper, and it would take several attempts to produce a preliminary layout that he/she would be happy with. The study stage is required to take the layout to the next level of design. The purpose of this stage is to establish:

- Elevations: equipment, piping, overhead clearances.
- Major support and column locations.
- Platforms, ladders, and ingress and egress requirements.
- Building sizes.
- Nozzle orientations.
- Equipment spacing between pumps, exchangers, vessels, etc.
- Control valve locations and meter run requirements.
- Bulk MTO.
- Start preliminary stress analysis, particularly in the racks to identify nested loops.
- Module limits.

Because it has a limited purpose and will not be used to generate construction drawings, the level of detail of the study model can be less than will be required for the detailed design model. The following are guidelines for study models:

- Model the NPS 4 and above piping only.
- Do not model pressure connections, temperature connections, or vents and drains.
- Specialty items, instruments, and vendor supplied packages that you have little information for can be modeled as simple blocks to identify the space requirement.

At the study stage all P&IDs, vendor equipment, and instrumentation are preliminary, and a considerable amount of manipulation of the model will be required if you maintain that the same models will be worked in continually from study through detailed design. Experience has shown that this can cause database corruption and rework, and will take longer than starting fresh, much the same as the manual drafters starting again with a clean sheet of Mylar® for the finished drawings. Therefore, new modelling should begin once the purposes of the study have been completed and firmer P&IDs and vendor data are available. The study models may be referenced into the detailed design models as a guide for detailed design.

The following two examples apply to the creation of study and detailed models on a project. These examples are per the CWP example and use modularized piperacks, an equipment module, and a field erected piping area.

4.4.1 Study Model Boundaries

Refer to Figure 4–2 Study Model Boundaries.

4.4.1.1 Notes

1. During the study stage all piping may be placed within one piperack model. During the study stage piperack module boundaries are established. Piperack modules will be remodeled as individual models during the detailed design stage.
2. Model boundaries are not always "straight lines" that cannot be transgressed. Logically, in this example, the expansion loop is in the piperack model.
3. During the study stage all piping may be placed within the field erected piping model. During the study stage, equipment module boundaries are established. Equipment modules will be remodeled as individual models during the detailed design stage.
4. For study purposes, piping leaving an area may be stopped at the model boundary. Note that this may not be the case for detailed design models.

4.4.2 Detailed Model Boundaries

Refer to Figure 4–3 Detailed Model Boundaries.

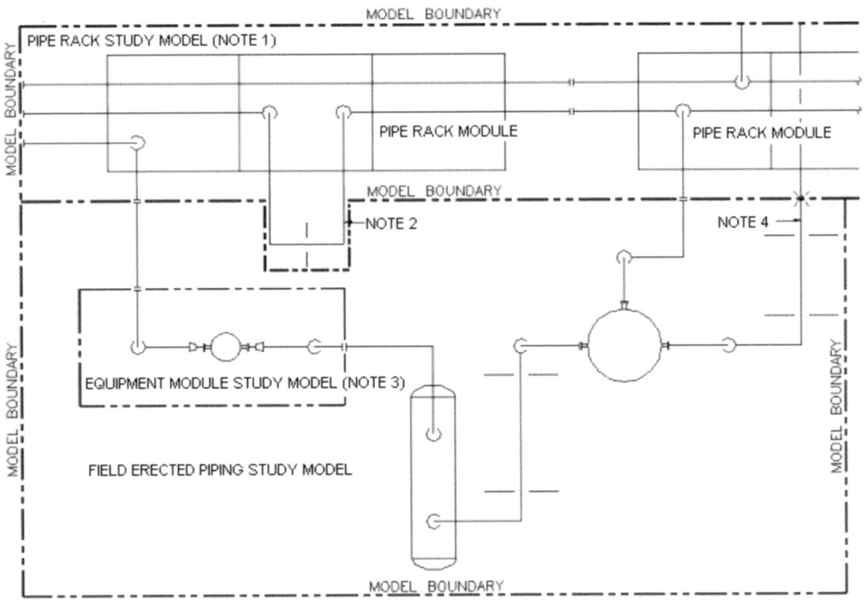

Figure 4–2 *Study Model Boundaries.*

4.4.2.1 Notes

1. Separate models are required for each piperack module.
2. Consider construction when dealing with piping and associated supporting. In this case, construction would require the loop spool in order to complete the piperacks, and therefore it is to be included in the piperack module model, not in the field erected piping model.
3. In this example the field erected piping CWP has been divided into two detailed design models.
4. Consider construction when dealing with piping that crosses the model and piping arrangement boundary. In this case, consider continuing the piping to a natural break such as a valve or the limit of the spool shipping box size. Breaking the pipe each time a model boundary is crossed introduces unnecessary field welding and added field costs. These small piping continuations are to be part of the model they originate in and included in the CWP.

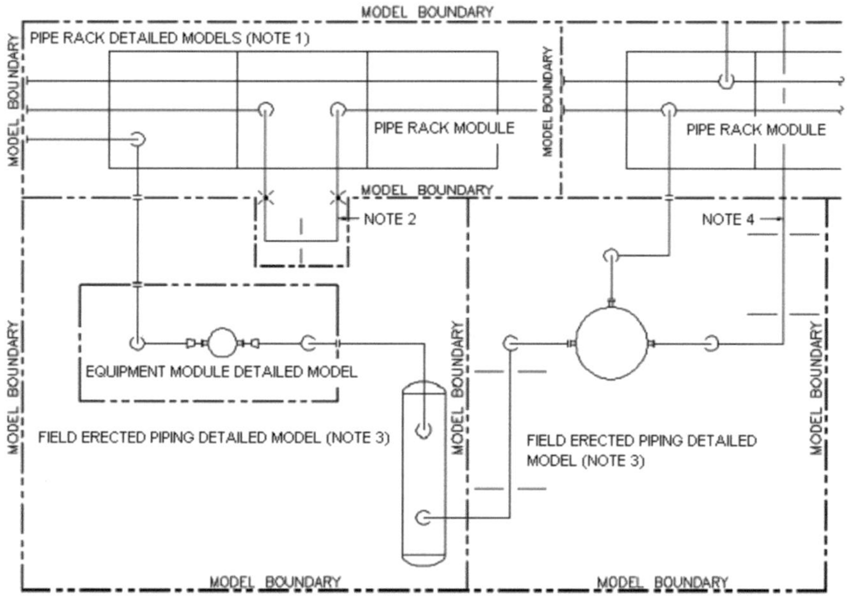

Figure 4–3 *Detailed Model Boundaries.*

4.5 Shop and Field Material Splits

The 3-D software will have settings for the "shop" and "field" material splits. Shop material is the material required by the spooling and module fabricators, and field material is the material required in the field for the erection and completion of the piping systems. You will have to inform your team of these material splits so that the designers make the correct selection when modeling.

These shop and field material splits will vary by industry, construction methodology, and the piping classes used. For example, in the oil industry piping classes often specify NPS 1½ and below piping as socket welded or threaded and NPS 2 and above piping as butt weld and flanged. The NPS 1½ and below piping is designated as field material because these are field run lines, while the NPS 2 and above piping is designated as a mix of shop and field. The butt welded fittings, weld neck flanges, and welded attachments are designated shop materials and the non-welded components, such as gaskets and bolts, are designated field materials required for the field erection.

The below example of shop and field material splits is based on oil projects that I have worked on. You will need to determine the material splits for the type of projects you are involved in.

- NPS 11/2 AND BELOW PIPING MATERIALS – FIELD-RUN, FIELD-ERECTED:
    - All materials, including pipe supports, are to be designated *"field,"* with the exception of TOL and SOL branches on NPS 2 and above piping which are to be designed *"shop."*
- NPS 2 AND ABOVE PIPING MATERIALS – SHOP SPOOLED, FIELD-ERECTED:
    - Bolts and Gaskets, Flanged Valves, Blind Flanges, Spectacle Blinds, Flanged Instruments, and Non-Butt Welded Specialty Items are to be designated *"field."*
    - Fittings, Welded Flanges, Butt Welded End Valves, Butt Welded End Specialty Items, and Butt Welded End Instruments are to be designated *"shop."*
    - Pipe supports are to be designated per Table 4–1 Pipe Support Shop/Field Designation Chart.
- EQUIPMENT AND PIPERACK MODULE PIPING MATERIALS – SHOP FABRICATED AND FIELD-ERECTED:
    - All materials, including pipe supports, are to be designated *"shop,"* with the exception of nuts, bolts, and gaskets at module connections which are to be designated *"field."*

Table 4–1 Pipe Support Shop/Field Designation Chart

Support Type	Shop	Field
Anchors (shipped loose, field installed)	X	—
Shoes (Welded shoes installed in shop. Clamp-on shoes shipped loose, field installed)	X	—
Base Supports, Dummy Legs, Trunnions	X	—
U Bolts	—	X
Guides (shipped loose, field installed)	X	—
Hangers & Springs	—	X
Repads	X	—
Field Supports	—	X

Correct designations are extremely important. Without accurate listings the MTO reports and the isometric BOMs will not have the correct information and purchasers will direct the materials to the wrong locations.

4.6 Procurement Splits

Establishing the above "shop" and "field" material splits will inform those dealing with the contracts and procurements as to where the piping materials must be directed, however, this does not clarify who is purchasing which of the materials. As a minimum, the engineering company will purchase all long delivery items such as high pressure and alloys pipe, fittings and valves, specialty items, and instruments, while the fabricators and erectors will purchase everything else.

Because there are many components in a piping system, statements, such as the last sentence above, do not give sufficient direction.

There are many ways for the material purchasing to be split and this may change from project to project. It is very important to document all of the possible piping components and who is purchasing which components in order to avoid any confusion. Needless to say, confusion over the purchasing of materials, overlooking the purchasing of materials, and misdirected materials all add up to schedule delays and added costs. To ensure that everyone has a clear understanding of the materials required and who is responsible for purchasing them, a summary document is required.

A good summary document is a piping material purchase matrix.

By far the best person to list the piping components and create this matrix will be you, the piping lead. You need to follow the contracting and procurement plans and ask questions in order to compile the piping material purchase matrix, however it is time well spent. This is a much needed document that will consolidate the information in one place and help to avoid potential material issues.

Refer to Table 4–2 Piping Material Purchase Matrix. This particular matrix is based on the engineering company purchasing all NPS 2 and above pipe, fittings, flanges, valves, and all specialty items and instruments, and the fabricators and erectors purchasing all NPS 1½ and below pipe, fittings, flanges, valves, and all pipe support materials.

The following sections are topics that have not been mentioned or discussed at any length to this point. They are in no particular order of importance, but are topics that must all be addressed early on and throughout the duration of the project.

Table 4–2 Piping Material Purchasing Matrix

	Field Erected Piping				Modules			
	SHOP SPOOLING FABRICATION (NPS 2 and above spooling and pre-fab of some field materials)		FIELD ERECTION (NPS 2 and above spools, NPS 1½ and below and other field materials)		PIPING FABRICATION AND MODULE ASSEMBLY (NPS all sizes)		FIELD ERECTION (NPS all sizes field material to connect module to module and module to off module)	
	Engineering Company Supply to Fabricator	Fabricator Supply	Engineering Company Supply to Erector	Field Erector Supply	Engineering Company Supply to Fabricator	Module Shop Supply	Engineering Company Supply to Erector	Field Erector Supply
PIPE MATERIALS , NPS 2 and above								
Pipe	X				X			
BW Fittings (tees, elbows, caps)	X				X			
WOL's	X				X			
Flanges (WN, SO, RJ, LJ)	X				X			
Orifice Flanges	X				X			
Blind Flanges			X		X			
Studs, Nuts, and Gaskets			X		X		X	
BW Valves	X				X			
Flanged Valves			X		X			
BW Specialty Items	X				X			
Flanged Specialty Items			X		X			
BW Instruments	X				X			
Flanged Instruments			X		X			
Lined Pipe and Fittings			X		X			
PIPE MATERIALS , NPS 1½ and below								
Pipe				X		X		
Nipples				X		X		
SW/Thr'd Fittings (tees, elbows, unions, couplings, caps, plugs)				X		X		
SOLs/TOLs		X				X		
Flanges (SW/Thr'd/Blind)				X		X		
Studs, Nuts, and Gaskets				X		X		X

Table 4–2 Piping Material Purchasing Matrix (cont'd)

	Field Erected Piping				Modules			
	SHOP SPOOLING FABRICATION (NPS 2 and above spooling and pre-fab of some field materials)		FIELD ERECTION (NPS 2 and above spools, NPS 1½ and below and other field materials)		PIPING FABRICATION AND MODULE ASSEMBLY (NPS all sizes)		FIELD ERECTION (NPS all sizes field material to connect module to module and module to off module)	
	Engineering Company Supply to Fabricator	Fabricator Supply	Engineering Company Supply to Erector	Field Erector Supply	Engineering Company Supply to Fabricator	Module Shop Supply	Engineering Company Supply to Erector	Field Erector Supply
SW/Thr'd Valves				X		X		
SW Specialty Items			X		X			
Flanged/Thr'd Specialty Items			X		X			
SW Instruments			X		X			
Thr'd Instruments			X		X			
Lined Pipe and Fittings			X		X			
Supports								
Welded Shoes		X				X		
Clamp-on Shoes		X				X		
Dummy Legs		X				X		
Trunnions		X				X		
Base Ell Supports		X				X		
Directional and Fixed Anchors		X				X		
Guides		X				X		
Hangers			X		X			
Spring Supports			X		X			
Field Supports (misc. structural, 'U' bolts, etc.)				X		X		
Miscellaneous Pipe Materials								
Re-pads		X				X		
Insulation				X		X		
Heat Tracing Manifolds		X						
Heat Tracing				X		X		

4.7 Issued For Bid and Bid Evaluations

During the course of detailed design you will be involved, at least to some extent, in issuing information for bid and the bid evaluations

4.7.1 Issued for Bid

Issuing documents for bid will take the form of reports and drawings. It is not always required to provide drawings, for instance, bids for bulk material purchases will be done using a bulk MTO report or a bulk valve report, or fabricators may base their bid on the reports of MTO, weld count, diameter inches of welding, and weights of materials. These reports are commonly compiled and generated by your material control and CAD support groups.

Should it be required to generate drawings to accompany a bid package then you must stamp them "FOR BID PURPOSES ONLY. NOT TO BE USED FOR CONSTRUCTION." You may be surprised at how often a fabricator will use these drawings after the contract is awarded to start the purchasing of materials and fabrication. At the least you will have done your part to stop this from happening.

Contracting strategies vary, but as an example, a spooling fabricator may receive a bid package for a CWP that includes:

- Isometrics.
- MTO report.
- Weld count report, e.g., 50 times NPS 12 welds.
- Unfactored diameter inches of welding report, e.g., an NPS 12 butt weld is 12 diameter inches of welding.
- Weights of materials.

With this information, plus other documentation such as specifications and standards, the spooling fabricator may price the job as follows:

- Unit prices for factored diameter inch welds. This price includes labor (cutting, bevelling, and welding) and handling charges (moving spools around, unloading materials, and loading spools). Factored inches are based on standard wall thickness, e.g., an NPS 12 standard wall thickness is .375" therefore an NPS 12 standard wall butt weld has a factor of 1 times the unit price (.375/.375) whereas an NPS 12 extra strong butt weld has a factor of 1.33 times the unit price (.500/.375).

- Unit prices for items such as shoes.
- The costs of the materials that are not supplied.
- Hydro-testing as an extra cost.
- X-ray and other NDE as an extra cost.
- Stress Relieving as an extra cost.

The above is only a broad overview and other factors may affect the bid price such as exotic alloys and spools that have two or more planes requiring extra set-up and handling time. Also, diameter inches of weld is only one method of pricing. Fabricators may use a price guide that lists the cost of welding standard wall carbon steel by the NPS size with factors for heavier walls and other materials. It is necessary that you understand the contracting and procurement strategies for your project.

4.7.2 Clarifications from Bidders and Bid Evaluations

You may be required to assist in the bid process by answering clarifications from the bidders and participating in bid evaluations from the point of view that everything has been covered. It is neither in the interest of the project nor the client to accept the lowest bidder, particularly when the bid is substantially lower than the others. This may be an indication that the fabricator has overlooked a major aspect of the work. It is not good to have to deal with a fabricator who is loosing money, or worse, on the brink of bankruptcy, in the middle of a project. High bids should also be looked at to understand if the bidder included work that the others did not. This could be a misunderstanding of the scope, but there is also a possibility that this bidder took into account scope missed by everyone else.

4.8 Equipment Coordinates and Elevations

The 3-D modelling software will be set up with reference points for equipment center lines and elevations, but in order for everyone to have a complete understanding of these, written documentation is required. The equipment center lines to be used for coordinates, and the reference points to be used for elevations, are required to ensure that everyone who is reading and marking up the equipment location plans has a clear understanding of the meaning of these coordinates and elevations, and that communication between the piping team and the other disciplines is clear. You should create a matrix for the types of equipment that will be encountered on your project similar to Table 4–3.

You should also agree the plant elevations to be used with your counterparts in the civil/structural group. This establishes further criteria that everyone will follow and eliminates a lot of discussion and coordination. The following are suggested based on a plant finished grade elevation of EL. 100.000:

- Equipment modules inside a building will be at TOS EL. AS SPECIFIED BY THE STRUCTURAL GROUP.
- Equipment modules outside of a building will be set at TOS EL. 100.600.
- Vertical vessels will be set at U/S B.P. EL. 100.300.
- Pumps will be set at U/S B.P. EL. 100.450.
- Other equipment will be set at U/S EL. AS SPECIFIED BY THE PROCESS/PIPING GROUP.

4.9 Module Design

Modularized design is the prefabricated assembly of structural steel skids and the associated piping, insulation, tracing and equipment at a fabrication shop for transportation to the field. As is the case with prefabricated pipe spools, there are many advantages to modularized design:

- Reduced field personnel and associated mobilization and camp costs.
- Shop labor costs are cheaper than field labor costs.
- The site work can proceed concurrently with the piping design.
- Large projects can use multiple fabrication shops.
- A shortened construction schedule.
- Better quality control due to a controlled environment and contented workers who live at home.
- More ground level access that reduces the need for scaffolding.

Table 4–3 Equipment Coordinates and Elevations Matrix

Equipment Type	Northing & Easting	Elevation
Vertical Vessel or Drum	Center Lines	Bottom Tangent Line
Horizontal Vessel or Drum	Horizontal Center Line and Center Line of Fixed End Saddle	Horizontal Center Line
Vertical Tank	Center Lines	U/S of Base Plate
Horizontal Centrifugal Pumps	Horizontal Center Line of Motor Shaft and Center Line of Discharge Nozzle	Center Line of Suction Nozzle or Motor Shaft
Inline & Vertical Pumps	Center Lines of Motor	Center line of Suction Nozzle
Shell and Tube Exchangers	Horizontal Center Line and Center Line of Fixed End Saddle	Horizontal Center Line
Aerial Coolers	Horizontal Center Line and Center Line of Inlet Nozzles	F.O.F. of Inlet Nozzle
Misc. Horizontal Equipment (Steam Generators, Air Make-Up Units)	Horizontal Center Line and Center Line of Major Feature (e.g., Stack)	U/S of Base Plate
Buildings	Center Lines of Building Column at North West Corner	H.P. of Finished Floor
Modules	Center Line of Steel at North West Corner	T.O.S.

The disadvantages are:

- Modules are limited in size by the transportation route: road access and weight limitations, telephone and power line clearance, tunnels and bridges.
- Extra structural steel required for rigging and bracing.
- Greater requirement for logistical planning and scheduling.

4.9.1 Design Considerations

You may have to make decisions on what will be modularized. Modularization lends itself well to equipment of a size that can fit on the modules within the shipping size, such as pumps and exchangers. Modularization also lends itself well to piperacks. The concept is to maximize shop installation of traditionally field-installed piping and peripherals, such as insulation and heat tracing, heat trace manifolds, sample coolers, utility stations, etc.

General design considerations for modules are:

- Transportation planning must happen early in the design stage since weight and size restrictions will determine the module parameters. Module sizes will vary according to transportation corridors and local and federal laws, but as an example may be in the range of 7.5m wide × 4.5m high × 30m long.
- Involve construction and operating personnel early in the planning and design stages.
- You may have to consider using a 1:40 scale if you are drawing the modules on ISO A1 or ANSI D size paper in order to show the complete module on one sheet.

Design considerations for equipment modules are:

- Revise existing layout and clearance specifications for lower space usage.
- Equipment and large valves subject to removal should be along the outer sides of the module for unobstructed access.
- Provide an interior access aisle 1m wide and 2.2m high.
- Valves and instruments should be accessible from the interior aisle or from the outside of the module, but should not project into either.
- Reserve space above the interior aisle for cable trays.
- Provide maintenance access around the module.
- Provide supply and return heat trace manifolds on the module in order that the tracing circuitry may be self contained, completed, and tested in the shop.
- Commonly, connections to the field erected piping are flanged, but may be butt welded.

Design considerations for piperack modules are:

- Flanged or butt welded connections may be used. When flanged connections are used it will be required to stagger the flanges at the module ends. For instance, assuming 6m bays, the centerline of support to FOF dimension for the pipe overhang will be alternately 2.7m and 3.3m.
- A horizontal dimension from centerline of support steel to the FOF or butt weld must be determined for the piping that leaves from the side of the module. A dimension of 500mm generally provides adequate clearance for bolt-ups or welding, and for installation of shoes.
- Expansion loops that extend beyond the module shipping width must be shipped loose and field installed.
- Lines exiting below the lowest piping level may have to be field installed if they drop below the bottom of the steel of the lowest module beam. The lowest module beam may be designed to accept the entire module load and to rest on the truck bed. Check with the structural engineer and the person arranging the shipping. It is possible to raise a module up on the truck bed by the use of shipping beams; however this may also reduce the allowable height of the module.
- Consider vents when determining the overall module height.
- A rule when deciding the elevation difference between piperack levels is 3 times the largest Nominal Pipe Size (NPS) of pipe that runs in the rack. Stringer levels, the steel beams between columns that support the pipes leaving the piperack, are set mid-way between the main beam levels or 1 ½ times the largest NPS pipe. This allows that the largest pipe can be rolled out of the rack with a fitting-to-fitting LR 90 degree elbow and a 45 degree elbow and that a pipe size of half the largest NPS pipe can use two fitting to fitting LR 90 degree elbows to leave the rack.

  For example, let us assume that the largest pipe size is NPS 24. An NPS 24 pipe is 610mm in diameter. Multiplying 610mm times 3 equals 1830mm between main beam levels and 1 ½ times 610mm equals 915mm to the stringers above and below the main beam levels. 915mm is also the center to center distance between two NPS 12 LR 90 degree elbows. Therefore all lines NPS 12 and below use two LR 90 degree elbows for elevation change to leave the piperack, and all lines NPS 14 up to NPS 24 will use one rolled LR 90 degree elbow and a 45 degree elbow.

If the largest pipe size is NPS 20 then the elevation difference between main beam levels will be 1520mm and 760mm between stringers and levels. The elevation change of 760mm will be by two LR 90 degree elbows for NPS 10 and below, and by a rolled LR 90 and 45 degree elbows for NPS 12 up to NPS 20. For NPS 16 the numbers become 1220mm, 610mm, NPS 8 and below and NPS 10 up to NPS 16 respectively. Given that there has to be space to work within the piperacks, 1220mm between levels is probably the minimum you would want to use.

- When laying out the piping for the larger NPS lines, make the 45 degree change in line with the pipe run whenever possible. There are two reasons for this: rolling the 45 degree change in the direction of the stringer may require a larger line space to the adjacent line, and can also result in having to push these larger lines further in towards the middle of the rack. Rolling towards the stringer may also create interference with the stringer before your elevation change can be accomplished.
- Decide the distance between bents. A common distance for supporting is 6m.
- Once you have established the elevation change between levels, stick with it throughout the plant. Running pipes from racks to sub-racks is made a lot easier when the stringer levels of the main piperacks match the levels of the sub-racks. Also, when you change direction, change the steel elevations by the same main beam to stringer dimension, don't turn flat.
- Legs will be required to elevate the piperack modules. The plant elevations of the piperacks are established by knowing the lowest allowed elevation above the roads and maintenance access requirements below the racks.

Much of the information above pertains to piperack layout in general whether they are modularized or stick built piperacks.

4.10 Module Numbering

It is a requirement that all modules have a unique number for tracking the designing, fabrication, transportation, and installation. You will need to establish the numbering system to be used for your modules. For this, look to your CWP numbering system. Use the unique CWP sequential number (usually a three or four-digit number). This number will require an abbreviated prefix to identify the module as an equipment module or a piperack module, such as EQM and PRM.

Transportation restrictions will put maximum limits on the width, height, and weight of modules for shipping. These restrictions mandate that breaks be incorporated into the designs. Therefore, it is common to have more than one module in a CWP. Modules that do not require vertical breaks, i.e., modules that have not been split into two or more stacked modules, can be identified by a numeric identifying suffix of 01, 02, etc., e.g., EQM-XXXX-01, EQM-XXXX-02, and PRM-XXXX-01, PRM-XXXX-02. Refer to Figure 4–4a Single Level Module Numbering.

Module designs that have a vertical split to accommodate the height restrictions of transportation require an additional identifier for the different levels. Two or more stacked modules can be identified by the addition of an alpha character to the suffix. The first alpha "A" identifies the module as being at the lowest level. Upper modules are sequentially B, C, etc., e.g., EQM-XXXX-01A, EQM-XXXX-01B, and PRM-XXXX-01A, PRM-XXXX-01B. Refer to Figure 4–4b Multiple Level Module Numbering.

You will need a set of rules for your area leads to follow when assigning the module sequential numbering. I have used the following guidelines on projects because it is a logical, easily remembered approach.

- Modules are numbered from west to east or from south to north with the exception that sub-piperack modules that branch from a main piperack are numbered from the main piperack outwards.
- For sub-piperacks that are in a different CWP to the main piperack, start with PRM-XXXX-01 for the first module (the one attached to main piperack) and number outwards.
- For sub-piperacks that are in the same CWP to the main piperack, start with the next sequential number for the first module attached to the main piperack, after numbering the main piperack modules. When several sub-piperacks are involved, start with the most westerly or southerly sub-piperack modules.

  This last rule is best explained with an example. Let us say that there is a main piperack running west to east comprising of ten modules with two branching sub-piperacks comprising of three modules each.

(a)

(b)

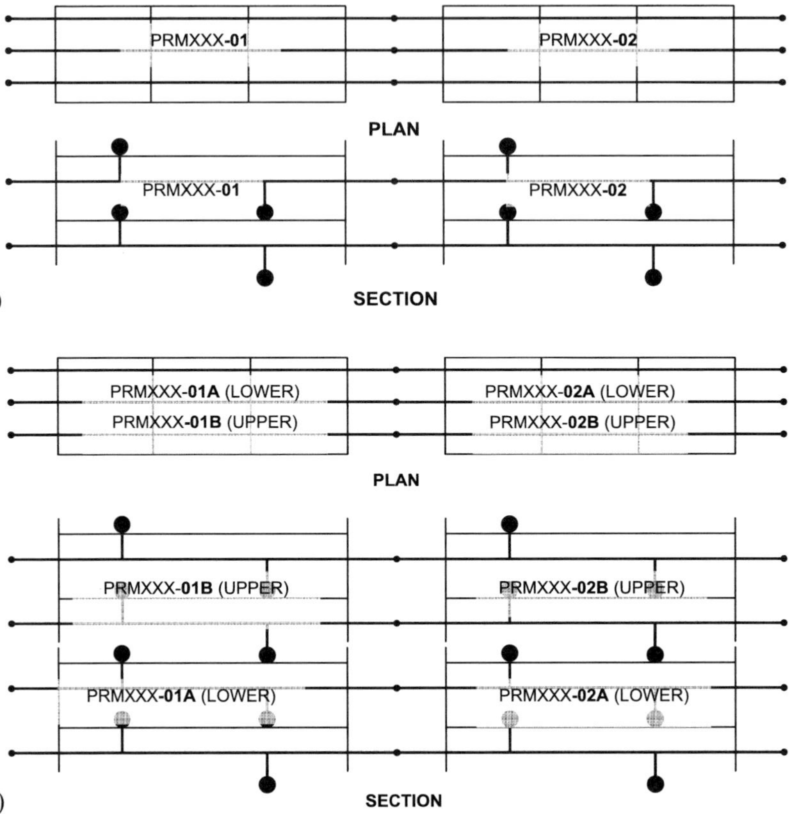

Figure 4–4 *(a) Single Level Module Numbering. (b) Multiple Level Module Numbering.*

> The main piperack is numbered PRM-XXXX-01 to PRM-XXXX-10 from west to east. The most westerly sub-piperack modules are numbered PRM-XXXX-11, 12, and 13. The next sub-piperack modules are numbered PRM-XXXX-14, 15, and 16.

It is important to have a set of rules in order to be consistent and avoid confusion. The above works well because everyone generally expects that numbering will go from west to east and south to north per the coordinate system. However, when dealing with sub-piperacks the expectation is that the numbering will go from the main piperack out as per the sequence that piperacks are erected—the main piperacks first, followed by the sub-piperacks working outwards from the main piperack.

4.11 Drafting Practice

You have a responsibility as the piping lead to ensure the quality of the deliverables that your group produces. One of your primary deliverables is the drawings, the quality of which are a product of the drafting practice. Drafting practice refers to the presentation and content of the drawings, and part of the lead's job is to ensure that the designers have adequate training and a vision of the finished product.

Regardless of the tools being used, what has not changed over the years is the requirement to portray the information in a clear and concise manner—the need to consider the layout, set-up, and content of the drawings. These are still human choices. The choices are just as important today as they were a hundred years ago for the end user, and the drawings are still a testament to the professionalism of the designer and the company as a whole. Drawings that are congested and poorly dimensioned are difficult to read and vastly diminished in their usefulness. Poor drawings can cause confusion leading to an increased number of Requests For Information (RFIs), lost productivity, and added schedule and costs.

4.11.1 A Brief History

In the days of manual drafting, which was common until very recently (many companies still employed manual drafting into the early 1990s), drafting practice or drawing presentation was one's calling card. Drafting practice referred to a nice hand lettering style, an ability to draw consistent line weights, and an ability to understand and accurately portray the information in a clear and concise manner. The manual drawing was not just pleasing to the eye, it was a clearly thought out document, and tremendous pride of ownership was taken in the presentation of the work.

With the advent of 2-D CAD the need to have a steady hand for line work and ability to hand letter became unnecessary as much of the presentation was dealt with by predetermined line weights, text styles, and symbols. The ability to move things around also allowed that some of the previously required advanced planning could be decided or adjusted later, something not possible for the manual drafter. The senior designer was now utilizing a new tool, the 2-D CAD software, but very little else changed in the approach to drawing creation.

Manual drafting and 2-D CAD drafting both require that the drawings are being created as the design progresses. The drafting of the plans and sections, details, dimensioning, and notes are an ongoing function of the design, decided by the senior designer during the project. It is this ongoing drawing development that differentiates manual and 2-D CAD drawings from 3-D CAD drawings.

4.11.2 3-D CAD Drafting Practice

Many projects now utilize 3-D CAD software, and as with 2-D CAD, the line weights, text styles, and symbols are predetermined. The drawings are generated after the design, in particular after the 3-D model, has been finished.

Drafting practice now includes management of the 3-D models and the databases to ensure the integrity of the information that later appears on the drawings and the reports that so many downstream people will later rely on.

To achieve this, the designer utilizing 3-D CAD has to manage a series of tasks never imagined by the manual drafters or even the 2-D CAD drafters. As the piping lead, you will require the full support of your CAD support group and CAD Job Notes (CJN) to provide training in order to ensure that these activities are carried out correctly. Many of these tasks must be carried out daily:

- Connectivity checking. The intelligent components can lose their connectivity to adjacent components in the model due to moving and copying. Models must have 100% connectivity to ensure trouble free generation of isometrics and electronic files that are to be imported into the stress analysis software.
- Database cleaning. When components are deleted from the model their associated data may be left in the database. It is also possible that items displayed in the model may no longer have associated data in the database. Therefore, it is required that the designers clean the databases of their piping models.
- Updating the components from the specification. The descriptive data in the model database files are generated as the model piping components are being created from the piping class material library. During the course of a project, the description within a piping class may change. For instance, a wall thickness. New components added to the model after this change will come with the new descriptor, but already modeled components will have to be updated for the new description to change accordingly in the isometric BOMs and material reports.

Note that these updates will only apply to the descriptions. Changes that affect the dimensions, such as a change of flange rating, will require that the previous components be deleted and will require that the designer remodel them.

- Running daily isometrics. To ensure that all models are functioning properly, designers should run an isometric of all added or modified lines toward the end of each day and correct any problems.
- Corrupted or disconnected databases. Occasionally, the problems will be beyond the knowledge of the designer to understand and correct. When this happens, the designer must stop all modeling and contact the CAD Support Group immediately. Continued modeling can cause permanent problems that can only be corrected by modeling everything over again.

In addition to the above daily tasks that must be carried out by the designers, the designers must also fix the issues that CAD support will bring forth in other reports. The CAD support group is not assigned to your project to respond only when there are problems. As is the case with any professionals that are expert in their field, the CAD support group will anticipate problems before they arise and will offer preventative maintenance throughout the project. Your CAD support will run a number of reports comparing models, databases and line designation tables to monitor the continued integrity of all aspects of the models. Discuss the frequency of these reports and how they will be addressed with the CAD support lead and your area leads. You may also have suggestions for reporting of your own that the CAD support will be able to implement.

The following are reports that you can expect from the CAD support group to be passed through you and your leads for investigation and action by the designers:

- Discrepancies between the designation of the components and the model number. The components must be designated according to the number of the model they reside in. A discrepancy indicates that the components are either incorrectly designated or that they are in the wrong model.
- The line numbers in the models match with the LDT. Throughout the project lines may be deleted or added, or line origins and terminations may change.

- The associated information for the line numbers in the models are consistent with the LDT:
  - The piping classes.
  - The NPS.
  - The process commodity abbreviation. Many companies have a designator for the process in their line numbering.
  - The insulation requirements.
  - The tracing requirements.
- Instrument Tags. A report can identify if instrument tags have the correct number of digits and prefixes, or that there are instruments without a tag associated with them.

Clearly, the designer of today utilizing 3-D CAD has a large responsibility and significant tasks to ensure trouble free drawing generation. But in many ways, the designers of today share the same challenges as their predecessors. Attention to detail and the task of producing a deliverable that is accurate and has clarity have been requirements for the vocation of piping designer throughout the years. The difference is that yesterday's designers were writing the information on their drawings, not typing the information into a database.

4.11.3 Ownership and Training

Successful model and database management alone will not guarantee successful deliverables. Poor drafting practice in terms of the portrayal of information may still present itself as a problem unless the lead is aware and takes the necessary steps to avoid this. Ownership of the models and drawings and training in drawing content are two other important aspects to consider. What used to be a natural progression and consequence of a younger junior designer moving through the ranks to that of a senior designer now has to be taught in a much shorter timeframe due to the following reasons:

- In the manual drafting days the leads and even the chief draftsman would walk around and observe the designs in progress. They received instant feedback by looking at the drawings on the drafting boards, and gave instant feedback on aspects of design and drafting to the designers. This form of mentoring was lost in the 2-D CAD world as it is not possible for senior personnel to easily follow the design on the designer's computer screen. He/she has to make a conscious effort to review the designs either on their own monitor or by making prints of the drawings.

- There is a lack of opportunity. Junior designers used to have the opportunity to hone their drafting and design skills by drafting P&IDs and standard drawings. They also learned from the more experienced designers by drawing isometrics from the finished piping arrangements, and assisting the senior designers in the drawing of sections and details. Once a junior had developed good drafting skills, they may have been trusted to backdraft the checked drawings for the senior designer. In short, they had many years of design and drafting involvement before ever reaching the goal of doing their own layout work. This opportunity has largely been lost in the 3-D world. Companies have little choice but to allow the junior designers to start designing in 3-D after a year or two of drafting P&IDs and standards, partly because there are no intermediate steps that can be taken and partly to keep them interested.
- An emphasis has developed on the utilization of the 3-D CAD software for designing, and a de-emphasis has developed on the drawings, particularly in the case of the isometrics, as an automatically or almost automatically generated product. It is now a widely accepted belief that the designer must work within the constraints of the software and that he/she is not responsible for the end result. The activity of drawing creation has been separated from the activity of designing and the designer no longer claims ownership of the finished drawing.

  The above lack of ownership may be compounded by coupling the assigning of the task of annotating the piping arrangements and generating the construction isometrics to the more junior personnel, and not to the designer. Unlike the juniors of the past who had the privilege of working along side the seniors, the juniors of today must generate drawings without knowledge of, or participation in, the design development. A disconnect often exists between the senior, the junior, and the end product.
- The drawings are no longer being worked on as the design evolves. They are generated toward the end of the design stage. At this point there is pressure to get the design finished and the drawings issued, which commonly translates into less time being spent on drawing consideration by the designers and the acceptance of some shortcomings in the finished product by the checkers.

You may think from the above that I lament the passing of the manual drafting era, and there is some truth to this. I miss the opportunity to admire the work of art of a skilled manual draftsman; it is a skill that I believe gave inspiration to the following younger generation of designers. I also believe that it is regrettable that the system of training and mentoring that prepared the junior piping designer to take on more responsibility as they progressed in their career has been lost.

However, I also believe in the efficiencies and benefits of CAD, 3-D CAD in particular, and I cannot imagine going back to the manual ways. It is a matter of recognizing how the design and drafting world has changed and adjusting accordingly. There are a few things you can do as the lead:

- Speak with the fabricators and constructors and find out what problems they may be encountering with the drawings and how things may be improved.
- Provide additional training and instruction in Piping Job Notes (PJNs) and CAD Job Notes (CJNs).
- Always have the designers generate their own piping arrangements and isometrics and do their own backdrafting to maintain ownership. Backdrafting in the 3-D world means making the changes in the models and regenerating the drawings.
- Never allow manually backdrafted edits to the drawings by anyone for any reason. The area designer must make all changes in their models, and the drawings are to be regenerated. This is not just a drawing ownership issue; it is also a model integrity issue.
- Create examples of the piping drawings as you wish to see them to give to your team. This will help to ensure a consistent product.

I have another reason as the lead for wanting the designers to generate and backdraft their own drawings. It is one of accountability and responsibility. I do not think it fair that someone else has to clean up the problems caused by poor attention to detail or the lack of understanding of another individual. Specifically, I am referring to poor modelling practices that can cause issues in reporting and drawing generation from the databases. By having each designer complete their area in its entirety I have the opportunity to gauge their capabilities and performance, and they have the opportunity to learn from their mistakes.

I realize that it is tempting and even unavoidable at times when an area is in a crunch to spread the work around, but do not do it as a general rule. Consider this during your manpower planning and scheduling.

The following sections discuss some areas of training and set-ups that I recommend.

4.11.4 Piping Arrangements

The two mistakes I see most often on piping arrangements are congestion and poor dimensioning. Piping arrangements require planning in order to avoid piping in the foreground that would block the detail of the piping in the background. Separating a plan into upper and lower elevations may be required to obtain the needed clarity. Careful attention must be paid to dimensioning. Over-dimensioning, double-dimensioning, dimensioning from matchlines, and dimensioning between pipes that are at different elevations are not useful practices.

4.11.5 Isometrics

The out-of-the-box product for automatic isometric generation requires set-up in order to achieve the desired line weights and format. The automatically generated isometrics will never look like their manually drawn counterparts, but they can be very functional. With a little effort and time spent with your CAD support group, a company configuration can be developed that will be very much to the satisfaction of yourself and the construction personnel.

Left to its own devices, automatic isometric generation will break unpredictably, and may have far too much information to be easily readable. Many lines may be crossing and smaller items, such as vents, often lack clarity because they appear too small on the drawing. Therefore, you need to have a few rules for your designers to follow:

- For shop fabricated piping, identify the field welds in the models. Field weld locations are decided by considering the shipping box dimension limitation and handling constraints such as floor, wall, and platform penetrations. (Note: identifying field welds is not required for modules.)
- Allow a maximum of 3 or 4 spools per isometric for clarity or the equivalent for module piping. Assign the isometric sheets and breaks in the model.
- Force the breaks between isometric sheets to be at a logical point, i.e., a flange or field weld location.

- The automatically generated isometrics want to break at line number changes. Include short branch connections as part of the main line isometric. The introduction of another isometric beginning with a short stub may be construed as a field weld location by the fabricators. For instance, imagine a header with a branch consisting of a WOL, pup piece, flange, valve, flange and branch line continuation. Should the branch line isometric begin right at the WOL, there is a good chance that the fabricator will assume a field weld location, whereas the WOL to pup piece can be a shop weld with the isometric break at the flange.

Consideration must also be given to the drawing numbering of the isometrics. It is very common practice to use the line number and sheet numbers for the isometrics of a given line. The problem is that it is impossible to schedule the generation of isometrics such that the sheets are numbered consecutively, starting with Sheet 1, from the beginning to end of the line in the direction of flow.

I suggest that you use a prefix to the isometric drawing number with such as the CWP sequential number/model number. By doing so, the isometrics for all lines within the CWP/model can begin with Sheet 1, and it is clear to everyone that the isometrics belong to that CWP. This can be a real advantage when there are thousands of isometrics to deal with.

For example, let us assume a CWP sequential number of 123, and two models numbered 123-01 and 123-02. Clearly, there will be other identifiers within the CWP and model numbers, the discipline being one of them for instance, but the CWP and model sequential numbers are the ones to focus on. Let us also assume a line number of HS-DVA-20-2210. The isometrics for model 123-01 would be numbered 123-01-HS-DVA-20-2210 SHT 01, 02, etc., and the isometrics for model 123-02 would be numbered 123-02-HS-DVA-20-2210 SHT 01, 02, etc.

4.12 Holds

It is not appropriate to hold back a CWP for issuing due to an inability to verify 100% of the information. It is better to release a package that is 95% complete for fabrication than to release nothing due to the 5% of unknowns.

During the checking stage there may still be some items for which the approved data are not yet available. An example can be a control valve face-to-face dimension. For these situations it is required that the checker create a Holds List and cloud the area of concern on the drawings with a "HOLD" cloud (the holds should also be listed on the drawing). This is a flag to the engineering and design teams that there is an unresolved issue still to be addressed, and it will notify the fabricator not to build this portion of the piping system until otherwise informed. The holds list should be compared against your needs list and the project action item list. Hold items that are not already identified should be added. In this way these holds will not fall through the cracks and be forgotten. Ideally you will be able to remove these and re-issue the drawings before the fabricator gets to them and is required to send an RFI. An RFI should prompt that the issue becomes a high priority. If it cannot be resolved, then a decision may be needed to proceed with the fabrication, possibly with field fit allowance added. This is a calculated decision that may later result in a further RFI during construction and an as-built to correct.

4.13 Project Binders and Lists

There are a number of project binders and lists that you must maintain throughout the project:

- Project binders.
- Piping Job Notes (PJNs).
- CAD Job Notes (CJNs).
- Action item list.
- Needs list.
- Equipment list.
- Data sheets.

4.13.1 Project Binders

You will need to create a master project binder(s) with the relevant piping information required by your designers. This will include:

- The Design Basis Memorandum (DBM).
- The construction, drafting, and piping execution plans from the Project Execution Plan (PEP).
- Piping specifications.
- Piping standards.

Have a copy made for each of the designers as they come on the project. As the project proceeds, you will need to keep the master up to date with the latest revisions and include the PJNs, CJNs, and other pertinent information being used on the project. It will be a valuable tool for orienting new members to your team.

4.13.2 Piping Job Notes and CAD Job Notes

Try as hard as we may, and for all the planning, there will be a need to issue instructions to your team of specific project requirements as the project unfolds. These will take two forms:

- Piping Job Notes (PJN) are instructions written by yourself or a designate on piping matters.
- CAD Job Notes (CJN) are instructions written by the CAD group on software related matters.

These should be numbered and recorded, e.g., PJN-001, 002, etc., with a revision number.

4.13.3 Action Item List and Needs List

You will need to create an action item list and a needs list. The action item list is a list of items and tasks within your control that require resolution, such as a CAD software issue.

The needs list is a list of all the issues that are holding up the progress, such as missing vendor data. These are tasks that will require the assistance of another group to resolve, sometimes with the influence of your project management team.

These lists should have several columns of information:

- An item number. 1, 2, 3, etc., will suffice.
- A brief description.
- The date of entry.
- The name of the person/group responsible for resolution.
- The date that the resolution/completion is required by.
- The date it was resolved.
- The priority, high, medium, low.
- The status, i.e., open or closed.

You will find that items for the action item list will end up on the needs list and visa-versa. This is not very important. What is important is that they get listed and resolved in a timely fashion.

Take these lists with you to your project meetings and discuss the items with the project team.

4.13.4 Equipment List

The equipment list is the responsibility of the mechanical engineering group. The document control group will keep a master expediting status report of the vendor information. You should keep your own up-to-date copy of the equipment list and receive the status reports for your records. Try to keep one step ahead and compare these with your schedule to determine if the vendor information is going to be available when your team will need it.

4.13.5 Data Sheets

There are several other documents to be maintained throughout the project for your team. These are:

- Spring supports list and data sheets. These will come to you from the stress group after the stress analysis for a given line has been completed.
- Specialty items list and data sheets. These will come to you from the mechanical/piping engineering group.
- Instrument list and data sheets. It goes without saying that these will come from your instrument group. You should request that a column be added to the instrument list for the identification of left and right hand mounts if one does not already exist. Ball pattern control valves in particular, and other types of instruments, can come in as either a left or right hand mount configuration, and as the designers are laying out an area they should fill this in to inform the instrument group of the correct orientation to be purchased.

4.14 Managing Standard Drawings

The standard drawings that are to be used on the project will be issued and frozen for the project at the beginning of the job. These standards may be from the client or your company, but because standards are ever evolving there will likely be revisions that will come your way. This is because the custodians of the standards are outside of the framework of the project. They will be listening to feedback from your project and other projects and revising and re-issuing accordingly. There are two reasons for issuing:

- A new standard has been created.
- An existing standard has been revised.

It is important that you have control of the acceptance of revisions to the standards and the issuing of new standards during the project. There are three reasons why a standard may be revised or a new standard created:

- Costs savings by changing the existing design.
- The existing design has to be changed due to a safety issue.
- A new standard is required to fill a gap.

These are all good reasons for issuing standards, but timing is everything and standards that do not emanate for known and approved project reasons should not be allowed on the project without the approval of the project manager. It is not likely that a safety issue would be turned down, but timing is important and other reasons for issuing may not warrant the repercussions they can have on cost and schedule. Do not forget that the project manager holds the purse strings to the project and must be made aware of changes to the standards.

Inform your team that if an unapproved standard comes their way they must bring it to your attention. Unapproved standards have been known to show up on a project and have led to the retrofitting of the already completed work. Should a new or revised standard be approved, be clear in your instruction of the usage, i.e., that it is only to be used as the project moves forward, not on already designed work, unless of course that is what you want.

4.15 Project Meetings

Like it or not you will have to attend project meetings with the project team and the other discipline leads, and have meetings with your own team.

The subject of the project team meetings will primarily be schedule. Are you on schedule, and if not, why not? This is where your action item list and needs list will be very relative. Get your concerns and issues out into the open and onto the project action item list. What you can do, and must do, is go into the project meetings prepared and armed with the information on what exactly is preventing you from meeting the schedule. What you cannot do is arrive uninformed and therefore vulnerable and unable to defend your position. The message you deliver may not be well received, but it will be respected if it is factual and accurate. If you come in with vague statements it will appear that you are not organized enough to get the job done, and others will doubt your abilities. Your project management team will usually be very good at their jobs, but they are not mind readers, and you have to be able to present your situation in a clear and decisive manner.

During the meeting do not make any commitments that you are unsure of. It is far better to walk away with the promise to investigate and report back than to commit in the moment under pressure and not to be able to deliver in the long run.

Your own team meetings will also be centered on the schedule. Listen, offer solutions, and document the issues. Here are a few guidelines for running your meetings:

- Have regularly scheduled meetings. Decide on the attendees and send out a recurring meeting notice. Primarily, the attendees will be your leads and the CAD manager, but also consider inviting the other discipline leads and your department manager as optional attendees. When staffing needs are going to be a topic, you should give your department manager the heads up and request that he/she be there. This is the primary person that will help you fill your staffing requirements.

  I find that one meeting every two weeks is sufficient. Weekly meetings come around too soon, although in the early stages when things may be progressing quite rapidly, you may want to start with weekly meetings and then switch to bi-weekly.
- Keep the meeting focused. People will often drift into tangents on topics that are best taken outside of the regular meeting. They will also start their own mini meetings while you are talking. Politely remind them that you are all there to discuss things as a group.
- Keep the meeting as short as possible. You can cover a lot in one to one and a half hours. Beyond this, people will start to become restless.
- Use the time to make announcements and share any news.

- Keep minutes. Distribute these and the updated action item and needs lists as soon as possible.
- Do not be late for your own meeting. If it is necessary to cancel a meeting be sure to send out a cancelation notice.
- I do not voice any rules for cell phones. No doubt we have all experienced cell phones ringing during a meeting, but I find that most people will turn them off immediately or excuse themselves and take the conversation outside of the room. Personally, I turn my cell phone off when I go into a meeting. The momentary disruptions that others' cell phones may cause have never warranted comment.

4.16 Progress Monitoring

Larger engineering companies will have software to track the progress of the piping design. The leads and designers must enter predetermined data into the required fields for progress monitoring. At smaller companies you will have to establish a tracking system. Discuss this with your project management team and your area leads to establish the parameters to be monitored. A simple spreadsheet divided by area will suffice, such as the examples Figure 4–5a Piping Design Progress Spreadsheet and Figure 4–5b Piping Checking Progress Spreadsheet.

It does not have to be anything fancy, in fact, you do not want to make it an arduous exercise, but it must have a breakdown that can reasonably be expected to be tracked, and to give a fair estimate of where you are in the progress. Referring to Figure 4–5a you will notice that the fifth column lists "Lines > or = to 4" modeled (15% of total effort.)." As long as you have the total line count of NPS 4 and above lines in Area 1, it is a fairly straightforward task to estimate the progress. Your designers should be "yellowing-off" their set of P&ID's and can tell you how many lines they have modeled. If a third of the lines are modeled, 5% would be entered.

It is a thankless task that few people want to do, but a very necessary one. I suggest that you have each area lead update the tracker for their area. Your project team will want this done on a regular basis, at least once a month.

From this, the earned value will be calculated which is a measure of how much real progress (i.e. value) has been achieved to date, and is the indication of how well you are doing.

Let us assume that a particular area has a total hour allotment of 5000 hours and the estimate is that you are 50% complete. You have earned 2500 hours. If you have actually used 2000 hours to this point

Construction Area	CWP/Model Area Number	Models included in Model Area	Equipment modelled to vendor data (0% to 3% of total effort.)	Lines > or = to 4" modelled (15% of total effort.)	Lines < or = to 3" modelled (10% of total effort.)	Lines > or = to 4" with stress comments incorporated (10% to 13% of total effort.)	Lines < or = to 3" with stress comments incorporated (7% of total effort.)	Valves modelled to vendor data (3% of total effort.)	Instruments modelled to vendor data (3% of total effort.)	Tempory Structural Steel modelled (7% of total effort.)	Clash Checking (final steel req'd) and Personal Isos run (7% of total effort.)	% Complete (ready for checking @ 60% to 65% complete)	Comments
Area 1	0001	5001		15%	8%	13%	7%	3%	3%	7%	7%	63%	1/ Miscellaneous piping to be completed.
	0002	5002		15%	8%	13%	7%	3%	3%	7%	7%	63%	1/ Miscellaneous piping to be completed.
	0006	1006		0%	0%	0%	0%	0%	0%	0%	0%	0%	1/ Not started. Require air make-up units
		8006	0%										
	0007	1007		15%	10%	9%	6%	3%	3%	7%	5%	58%	1/ Require info on OTSG to complete stress. 2/ Miscellaneous piping to be completed.
		8007A	2%										
		8007B	2%										
	0010	1010		15%	10%	9%	6%	3%	3%	7%	5%	58%	1/ Require info on OTSG to complete stress. 2/ Miscellaneous piping to be completed.
		8010A	2%										
		8010B	2%										
	0011	1011		15%	8%	10%	5%	3%	3%	7%	5%	56%	1/ Missing final vendor data. 2/ Miscellaneous piping to be completed.
		8011	3%										
	0013	1013		15%	10%	9%	6%	3%	3%	7%	5%	58%	1/ Require pre-heaters. 2/ Require info on OTSG to complete stress. 3/ Miscellaneous piping to be completed.
		8013A	2%										
		8013B	2%										
	0014	1014		15%	8%	10%	6%	3%	3%	7%	6%	58%	1/ Miscellaneous piping to be completed.
		8014	3%										
	0015	1015		0%	0%	0%	0%	0%	0%	0%	0%	0%	1/ Not started. Require air make-up units
		8015	0%										
	0016	1016		13%	8%	8%	5%	3%	3%	6%	4%	50%	1/ Require air make-up units 2/ Require pre-heaters. 3/ Require info on OTSG to complete stress. 4/ Miscellaneous piping to be completed.
		8016A	2%										
		8016B	2%										
	0017	1017		8%	5%	5%	3%	2%	2%	4%	6%	60%	1/ Catalog info only on glycol circulation pump. 2/ Utility piping to be completed.
		6017		8%	5%	5%	3%	2%	2%	4%			
		8017	3%										
	0018	1018		15%	10%	10%	6%	3%	3%	7%	6%	60%	
		8018	3%										

Figure 4–5a *Piping Design Progress Spreadsheet.*

Construction Area	CWP/Model Area Number	Models included in Model Area	Ready for checking (60% to 65% of total effort.)	Equipment checked in model (0% to 2% of total effort.)	Piping Arrangements generated (5% to 10% of total effort.)	Isometrics generated (5% of total effort.)	Piping Arrangements checked (15% to 20% of total effort.)	Isometrics checked (0% to 3% of total effort.)	Piping Arrangements backdrafted (3% to 5% of total effort.)	Isometrics backdrafted (0% to 2% of total effort.)	Package IFC	Comments
Area 1	0001	5001	63%		10%		20%		5%		98%	Issued with some "HOLDS".
	0002	5002	63%		10%		20%		5%		98%	Issued with some "HOLDS".
	0006	1006	0%		0%	0%	0%	0%	0%	0%	0%	
		8006		0%								
	0007	1007	58%		0%	0%	0%	0%	0%	0%	58%	
		8007A		0%								
		8007B		0%								
	0010	1010	58%		0%	0%	0%	0%	0%	0%	58%	
		8010A		0%								
		8010B		0%								
	0011	1011	56%		0%	0%	0%	0%	0%	0%	56%	
		8011		0%								
	0013	1013	58%		0%	0%	0%	0%	0%	0%	58%	
		8013A		0%								
		8013B		0%								
	0014	1014	58%		0%	0%	0%	0%	0%	0%	58%	
		8014		0%								
	0015	1015	0%		0%	0%	0%	0%	0%	0%	0%	
		8015		0%								
	0016	1016	50%		0%	0%	0%	0%	0%	0%	50%	
		8016A										
		8016B		0%								
	0017	1017	60%		2%	2%	0%	0%	0%	0%	68%	
		6017			3%	3%	0%	0%	0%	0%		
		8017		0%								
	0018	1018	60%		0%	0%	0%	0%	0%	0%	60%	
		8018		0%								

Figure 4–5b *Piping Checking Progress Spreadsheet.*

(hours being used are referred to as the burn rate) then your productivity is the earned hours divided by the actual hours, which is 2500/2000 or 1.25. This would be amazing and likely an indication that the progress is being overestimated, that the budgeted hours are overestimated, or that your designers are charging some of this work to another code.

On the other hand, if you used 3000 hours then your productivity is 2500/3000 or 0.83, which is an indication that you are in danger of going over the budgeted hours and schedule. One month of low productivity may raise some eyebrows and questions, but would not be of great concern in the big picture. Continued low productivity will prompt investigation to determine what is happening. Possibly the budgeted hours are underestimated, possibly some trending has been overlooked, or it may be that your designers are charging other work to the code for this work.

For that matter, continued high productivity will be investigated to determine the truth of the reporting. Is there overestimating of the productivity happening or were the hours originally overestimated? A productivity of 1 is, of course, right on target.

These reports will be used to forecast the burn rate, the completion dates against the original target dates, and will help determine where some overtime may be required or resources may be re-allocated to bring things back on track.

They will also be used to forecast the final costs, i.e., if you are going to go over budget. Clients may also only be willing to pay based on the earned hours and not on the actual hours used.

Progress monitoring is useful in estimating how the project is likely to finish in terms of budget and schedule, and serves the very important purpose of identifying when you are running into trouble. It also helps to identify the requirement for a mid-course adjustment before you are so far into trouble that you can never recover.

4.17 Design Change Notice (DCN)

A Design Change Notice (DCN) is a document distributed from the project team to the affected disciplines to give notification of an approved change to the design scope or deviations affecting the budget. This document includes:

- The details of the change, with a sketch if needed.
- The reason for the change.
- The impact on the schedule.

You must distribute these to your leads as soon as you receive them. A DCN often follows an approved scope change request, but can also be used to document design changes that are not scope changes but are changes in budget. For instance, a deviation from the material specified in a piping class to a substitute material is not a scope change, but can affect cost. Also be aware that a DCN may trigger a trend.

4.18 Field Change Notice (FCN)

A Field Change Notice (FCN) is a document used to communicate change to the fabricators and construction after the drawings are issued IFC, and to the affected disciplines. This can be changes in scope, changes in design, and the adding or removing of holds. As with the DCN, this document is distributed from the project team to the affected disciplines and similarly includes details of the change, the reason for the change, and the impact on the schedule.

For any changes that you know are going to happen, you have to mark-up the affected drawing(s), then pass this on to the project engineer to notify the field or the fabricator immediately that this change is coming. The project engineer will issue the FCN to officially document the revision and make sure all disciplines involved are aware of the change. Following this you must revise and re-issue the drawings.

4.19 Request For Information (RFI)

A Request For Information (RFI) is a document submitted by the fabricators or the construction team and can be for any number of reasons, ranging from clarifications to the specifications, standards and IFC drawings, to material deliveries and transportation requirements.

The RFIs that are designated for action by you must be answered as soon as possible. Your response must be a clarification to the question asked and may have to include mark-ups to drawings that may become revisions to these same drawings.

An RFI may trigger a trend or scope change and a DCN or FCN.

Whether the document submitted to you is a DCN an FCN or an RFI, this is where your stick files will be of great value. Note the changes on the P&ID, piping arrangements and isometrics stick files, with the number of the DCN, FCN or RFI. Revise and re-issue the drawings in a timely fashion.

Going back to the need to maintain project binders, these documents are no different, and you must maintain a log of each in a binder for future reference.

CHAPTER 5

Shop Fabrication

5.1 Introduction

During the course of the project a shop fabricator or fabricators will be engaged for pipe spooling and, if also utilized on the project, module fabrication and assembly.

At many companies and on many projects a close working relationship between the piping lead and the fabricators is a given responsibility of the lead, but not every project manager requires the lead's services aside from the contributions you make to the Scopes of Work (SOW) and responding to Requests For Information (RFIs). Therefore, you will need to discuss the level of involvement that you will have with the fabricators with each project manager and his/her expectations of you on a project by project basis.

In this chapter I will discuss the benefits to the project when a close working relationship exists between the piping lead and the fabricators and the role of the lead in this process. Working closely with the fabricators is required of all piping leads sooner or later.

You should build a rapport with all of the fabricators, and if possible visit with them on a regular basis. A close working relationship will pay dividends to the project in terms of resolving problems and addressing potential problems.

As the piping lead, you will interact with and provide a number of services for the project management team in support of fabrication:

- Attend the kick-off meetings.
- Provide fabrication and assembly Scopes of Work (SOW).
- Provide instruction to the fabricators.

- Respond to Requests For Information (RFI).
- Visit with the fabricators.

Prior to the contract awards and these meetings you will be required to have input into the piping SOWs, if not write them entirely. You should meet with the construction manager and discuss his or her requirements for shop fabrication drawings, and spool identification and module identification markings.

Based on your meeting with the construction manager, you should write an instruction for the spooling fabricator on the drafting aspects of the scope as a supplement to the SOW and explain what will be provided to them and what will be required from them. This will cover the drafting of spool sheets and will contain information not covered in the SOW. Later in this chapter is an example of an instruction given to the spooling fabricator.

5.2 Kick-off Meetings

Kick-off meetings will be conducted with every successful fabricator prior to the commencement of work. These meetings introduce the fabricator to the project, the organization, the key contacts, the scope of the work, the expectations, and also allows for questions and clarifications. Generally these meetings will cover:

- Introductions.
- Content of the Construction Work Packages (CWP).
- The Scope of Work (SOW).
- Materials and services, i.e., client supplied and fabricator supplied.
- Quality Assurance and Quality Control (QA/QC) requirements.
- Inspection requirements.
- The schedule.
- The Request For Information (RFI) procedure.
- Report requirements.
- Contact information.
- Expediting.
- Contracts and administration.

The common representation from your company will be the project engineer, QA/QC coordinator, contracts administrator, purchasing agent, construction manager, and shop inspector. Your company is the client to a fabricator, and his/her manager, a project engineer, and other senior management types will commonly represent the fabricator's company.

It is not a sure thing that a fabricator will bring his/her drafting manager. The drafting manager is the person responsible for receiving your drawings, interpreting them, and creating the shop fabrication drawings, and consequently you will need to speak with him/her. Discuss the importance of the drafting manager of each fabricator being present at these meetings with your project manager and have him/her request their presence.

In the meetings with the fabricators you will need to explain any instruction and establish their drawing and model requirements:

- What CAD system are they using? Some fabricators will have the ability to work with imported files from the 3-D models to automatically generate spool sheets, just as your piping department have the ability to automatically generate isometrics.
- If the fabricator cannot work with the imported model files, they may still be interested in receiving the models in a viewing file format for planning and information purposes.
- Do they have the capability to conduct a model walk-through? That is, do they have viewing software and a conference room set-up with an overhead projector? This is useful for reviewing the CWP models when you visit.
- Should the fabricator have the ability to work with the imported files and/or the models, you will need to secure the name of the individual that will be responsible for receiving them. You will have to establish contact between that individual and your own CAD support manager for issues surrounding the imported files and viewable 3-D model files. Make sure that you will be copied on all correspondence between them.
- Confirm the fabricator's requirements of size and the number of copies of the drawing packages. The ANSI B or ISO A3 reduced size copies of the piping arrangements are sufficient for reference, but many fabricators will want full size ANSI D or ISO A1 size copies for the shop to work with.

5.3 Scopes of Work (SOW)

The SOW must describe in detail the work that the fabricator is to do. It must also describe the work that is not within the fabricator's scope. It is apparent that the contents of a document of this nature cannot be decided at this point in the project. It is also apparent that in order to write this document you must have intimate knowledge of the construction and purchasing strategies. It is for these reasons that in the previous chapters we have discussed the importance of setting up and running your project according to the project execution plan, the construction execution plan, construction work package boundaries, model boundaries, drawing boundaries, and shop and field material splits.

The SOW is the culmination of the work you have been planning and executing throughout the project. The contents of the SOW come from decisions that were made at the beginning and throughout the project. It is most likely that your company will have templates for SOWs that have been used on previous projects that you will be able to modify to suit the requirements of your project.

The following are two examples of SOWs. One is for pipe spool fabrication and one is for module fabrication and assembly. You will notice that they reflect the same shop and field material splits and fabrication splits as discussed in Chapter 4. They also discuss who is buying which materials. The projects you are involved in may or may not follow the same parameters.

5.3.1 Example of Pipe Spool Fabrication Scope of Work

1. Scope of Work:

   Work to be performed under this Construction Work Package (CWP) No. ________________.

   Work includes the following items, all as set forth in the drawings, specifications, and other data:

   - Fabricate welded spools of piping components NPS 2 and above in accordance with the attached drawings, line designation tables, piping standards, and specifications.
   - Fabrication dimensions are to be taken from the isometrics.
   - Field welds (FW) are indicated on the isometrics. Spools are to be to a maximum shipping box size of 3mH × 3mW

× 12mL + or – 300mm as marked-up on the isometrics. Should the contractor encounter field weld positioning on the isometric that would result in a spool size beyond these parameters, an RFI must be raised.

- Additional field welds that the contractor may deem necessary must be approved first by way of an RFI.
- Field welds have no trim allowance. Fabricate spools to exact dimensions.
- Contractor may assume his/her standard weld gap when calculating the cut length of pipe ending in a FW. A BE preparation is required.
- Field fit-up welds (FFW) are to have 150mm of additional pipe length for trim allowance. Finished BE preparation is not required.
- Fabricate and install the welded shoes.
- Fabricate clamped on shoes, guides, and anchors. Ship loose.
- Provide a stub for the base ell supports per the standard drawings. Fabricate the remainder of the support per the standard drawings. Ship loose.
- Fabricate the dummy legs, trunnions, and re-pads. These are welded attachments to the spools.
- Pipe to be painted according to the line designation table requirements.
- Each shipment must be accompanied by a list of the spools and copies of the spool sheets.
- The spool sheets must be signed-off copies indicating completion of each stage of fabrication and testing. All spool sheets are to include the material heat numbers and the weld mapping.
- Spools must have an aluminum tag (sized 25mm × 75mm) tightly wired with SS wire to them with the mark number stamped into the tag. As an additional precaution the mark number should appear on carbon steel spools by way of paint pen writing, and stainless steel spools by way of an approved marker (felt style marker with less than 50 ppm chloride content).
- Shipped loose shoes, guides, and anchors are to be identified by the standard tag number by paint pen writing.

- All documentation to be provided with piping material when released to site by the contractor.
- Method for the material requisitions by the contractor to be determined by contract administration in conjunction with the contractor.
- Excess materials to be shipped to site as determined by contract administration in conjunction with the contractor.
- Turnover package to be provided in whole and as defined by the QA/QC requirements.
- For shipping, arrangements are to be made between contract administration and the contractor.
- All ends of pipe and flanges must be protected for shipping.

2. Work Not Included:

   The following items of work associated with this CWP do not form part of the Work of the CWP:

   - Hydro-testing.
   - Assembly and erection of spools and non-welded attachments.
   - Insulation and tracing.
   - Fabrication of heat tracing manifolds.

3. Supplied Materials:

   The following materials will be supplied for incorporation into the work:

   NPS 2 and above:

   - Pipe
   - BW fittings (tees, elbows, caps)
   - WOLs
   - Flanges (WN, SO)
   - BW valves
   - BW specialty items
   - BW instruments

Except for those items specifically listed above, the contractor will be responsible for supplying all materials required to complete the work of this CWP.

4. Schedule:

   Work of this CWP shall be performed in accordance with the following schedule:

   - Work shall commence on or about _______________.
   - Work shall be completed on or about _____________.

5. Drawings, Specifications, and Other Data:

   Work shall be performed in accordance with the drawings (signed and stamped "Issued for Construction"), specifications and other data attached hereto and listed below:

A pipe spool is a portion of a piping system limited in its dimensions for ease of handling and shipping. The spools consist of the welded components in the piping system with naturally occurring breaks at the flanges and introduced breaks at fittings in the form of field welds. Figure 5–1 is a photograph of an individual spool in a pipe fabrication shop and Figure 5–2 is a photograph of multiple spools on a trailer loaded for shipment from the fabrication shop to the field.

5.3.2 Example of Module Fabrication and Assembly Scope of Work

1. Scope of Work:

   Work to be performed under this Construction Work Package (CWP) No. ___________________.

   Work includes the following items, all as set forth in the drawings, specifications and other data:

   - Complete fabrication and assembly of the piping systems and structural frames in accordance with the attached drawings, line designation tables, standards, and specifications.

Figure 5–1 *Photograph of Finished Spool.* (Printed with the permission of Pepco Pipe Services, Edmonton, Canada.)

Figure 5–2 *Photograph of Spools Loaded for Shipment.* (Printed with the permission of Pepco Pipe Services, Edmonton, Canada.)

- Spools that cannot be installed, e.g., expansion loops that would bring the module beyond the allowable shipping width, are to be shipped loose attached to the module.
- Installation dimensions are to be taken from the piping and structural general arrangements and isometrics.
- Verify shim height requirements as noted and install.
- Delicate instruments, as identified in attached documentation, are for pre-assembly only. These are to be removed and crated for shipment to site. Fill the gap with a temporary spool piece for module shipment.
- Should some bolted-in items not arrive in time for shipment, at the discretion of the contract administration, the contractor may be instructed to ship without these items. In this event fill the gap with a temporary spool piece for module shipment.
- Install pipe hangers and spring supports. Spring supports are to remain locked for shipping.
- Hydro-test per specifications and line designation tables.
- Pipe to be painted according to the line designation table requirements and attached specifications.
- Heat trace per the attached drawings, standards, and specifications.
- Insulate per the attached drawings, standards, and specifications.
- Prior to shipping, a final check must be conducted to verify the locations of the connections at the module edge. Should any connections differ in location from the dimensions as noted on the piping arrangements, an RFI must be raised. A decision will be made to either:

  Leave them as fabricated. Adjustments will be made to the connecting piping in the field.

  Adjust in the shop to avoid later field connection problems.

- The module steel must be marked at the base with the module number by way of 100mm high white paint lettering using a stencil. The "DATUM POINT," as shown on the piping and structural general arrangements, must also be marked clearly on the module base.

- Contractors who create spool sheets must provide these with the final turnover documentation. Contractors who work to the isometrics must mark-up a set of isometrics with any as-builting, material heat numbers, and weld mapping.
- Method for material requisitions by the contractor to be determined by contract administration in conjunction with the contractor.
- Excess materials to be shipped to site as determined by contract administration in conjunction with the contractor.
- Turnover package to be provided in whole and as defined by the QA/QC requirements.
- For shipping, arrangements are to be made between contract administration and the contractor.
- Temporary strapping, hold-downs, and bracing for shipping are to be determined by the client inspector in conjunction with the contractor. These are to be painted bright red for easy identification and removal at site.
- All ends of pipe and flanges must be protected for shipping.

2. Work Not Included:

   The following items of work associated with this CWP do not form part of the Work of the CWP:

   - Final erection at site.

3. Supplied Materials:

   The following materials will be supplied for incorporation into the work:

   NPS 2 and above:

   - Pipe.
   - Fittings (tees, elbows, caps).
   - WOLs.
   - Flanges (WN, SO).
   - Valves.
   - Studs, nuts, and gaskets.

All:

- Specialty items.
- Instruments.
- Pipe hangers.
- Spring supports.
- Structural steel members.

Except for those items specifically listed above, the contractor will be responsible for supplying all materials required to complete the work of this CWP.

4. Schedule:

   Work of this CWP shall be performed in accordance with the following schedule:

   - Work shall commence on or about ______________.
   - Work shall be completed on or about ____________.

5. Drawings, Specifications and Other Data:

   Work shall be performed in accordance with the drawings (signed and stamped "Issued for Construction"), specifications and other data attached hereto and listed below:

A module is a completely assembled and tested unit consisting of all piping, equipment, steel and electrical components allowed within a predetermined size. A module is built in a module fabrication shop and is shipped to the field for final installation. The dimensions of a module are limited by the type of transportation, e.g. truck, rail, barge, and the height, width, length and weight restrictions imposed by the transportation corridor and jurisdictions that the module will pass through. Figure 5–3 is a photograph of a piperack module being loaded for shipment. Figure 5–4 is a photograph of a piperack module starting the journey from the module shop to the field. Figure 5–5 is a photograph an equipment module ready for shipping.

5.4 Instruction to the Fabricator

Scopes of Work outline specifically the 'what' aspect of the work. There are times when it is necessary to outline the 'how' aspect of the work. Each fabricator will have his/her own procedures, and while the

Figure 5–3 *Photograph of a Piperack Module being loaded for shipment. Notice the temporary shipping steel at the front of the module.*

Figure 5–4 *Photograph of a Piperack Module being shipped. Notice the fire proofing and shipping beams to raise the module steel off of the truck bed due to the piping that exits below the lowest level of the module.*

Figure 5–5 *Photograph of an Equipment Module ready for shipment.*

outcome may be very similar, the differences could cause some difficulties for construction. It would not be efficient to have different approaches to the mark numbering, the spool sheet drawing content, and the spool drawing package assembly. When you are dealing with multiple fabricators it is best to assert control over these details and provide guidance. The written instruction to the spooling fabricator is to ensure consistency.

Instruction of this nature is not commonly required for the module fabricators because their scope includes complete fabrication and assembly, whereas the spools are often handed over to a different contractor for erection.

As with the SOWs, the following example reflects the same shop and field material splits and fabrication splits as discussed in Chapter 4.

5.4.1 Example of Instruction to Fabricator

1. Introduction:

   This document is to inform pipe spool fabricators of the additional requirements beyond the Scope of Work (SOW) for spool sheet drafting, and to provide information on the drawing packages and the engineered piping drawings.

It gives instruction and guidance on expectations to ensure a consistent deliverable for QA/QC and field construction. While every attempt has been made to provide clarity to fabricators, it is likely that the need for further questions or clarifications may arise. Questions and clarifications will be answered by way of an RFI. In the event that the information contained herein differs from the instruction in the SOW, then the latter shall govern.

2. Engineered Drawings:

   Piping arrangements and isometrics are issued in the Construction Work Packages (CWP). Each CWP drawing package comes with a transmittal, a SOW, a drawing cover sheet and a drawing index. Fabricators are responsible for reviewing the drawings provided against the transmittal and the drawing index to ensure that they have all the drawings at the listed revisions.

   Issues arising during fabrication that require a change or deviation from the design must first be cleared by way of an RFI. Note that a CWP contains a complete set of all drawings required for the entire fabrication and erection of the piping systems, and may well contain drawings that are not part of the pipe spool fabricator's SOW. Primarily this would be NPS 1½ and below socket weld and threaded piping.

   - There are piping plans only. No sections.
   - Piping arrangements have minimal dimensioning for reference, and are to be read in conjunction with the isometrics which are the governing documents.
   - Fabrication dimensions are to be taken from the isometrics.
   - Material is listed in the bill of material on the right-hand side of the isometric.
   - Materials are listed as "SHOP" or "FIELD." These are a construction split, not a material purchase split. SHOP materials are those required in the shop. Some of the shop materials are client supplied and some are fabricator supplied. FIELD materials are those directed straight to the field for later completion/erection of the piping system.
   - For complete instruction on client supplied and fabricator supplied materials see the SOW.

- The isometrics indicate the required field weld (FW) and field fit-up weld (FFW) locations.

3. Spool Sheets (Fabrication Drawings):

- Spooling only applies to NPS 2 and above butt-welded piping. The components of a spool are the butt-welded fittings and flanges, welded on branch connections, e.g., WOL, TOL and SOL, and welded on attachments, e.g., dummy legs and shoes.
- Other than that noted in the following two bullets and within the SOW, ignore any isometrics of NPS 1½ and below piping.
- All NPS 1½ and below SOL and TOL branch connections are full penetration weld attachments to the NPS 2 and larger spools, and as such are considered part of the shop spooling.
- All swages NPS 2 and above times NPS 1½ and below require a butt weld at the large bore end, and as such are considered part of the shop spooling.
- It is not allowed to have multiple spools detailed on one drawing. Each spool must have an individual spool sheet.
- The mark number must appear in the drawing title block. Each spool sheet must have a revision number starting at Rev. 0.
- The drawing number and revision number of the isometric that the spool sheet has been drawn from must appear as a reference within the title block.
- The project name must appear in the title block.
- A complete bill of material and heat numbers must appear on the right-hand side of the drawing.
- A "North Arrow" must appear in the upper left-hand corner of the drawing. Spool orientation and the north arrow (commonly up and to the left) must be drawn to the same orientation as the isometric.
- All components must be shown on the spool sheet and accurately located by a dimension from a reference point, e.g., from the centerline of an elbow or tee, or from face-of-flange. This includes all items that are welded attachments such as shoes, and shipped loose items such as guides and anchors. Do not show items that are not

part of the spooling SOW, e.g., bolts, gaskets, flanged valves, and flanged instruments.

- Shipped loose items (e.g., anchors, guides, base ell supports) will be identified with a leader arrow and the notation "FIELD WELD."
- Drawing continuation references to the adjoining spool sheets or to equipment are required.
- All QA/QC information, as has been determined by the QA/QC coordinator, must appear within the title block of the spool sheet.

4. Spool Mark Numbering:
   - The spool mark numbering that is to be used by all fabricators is indicated on the isometrics. A control number may be assigned by the fabricator, but must be marked-up on the close-out set of isometrics.
   - Should a spool be added, as will be the case when an additional FW is added, the fabricator is to re-number the mark numbers on the isometric as an as-built.
5. Shipping:
   - Provide a list of the spools and a copy of each spool sheet of the spools being shipped.
   - Shipped loose items will carry the same mark number as the spool they are associated with, i.e., the spool sheet they appear on.
   - In addition to the other required QA/QC documentation, at the completion of each CWP the close-out must be accompanied by a spool sheet drawing list, the spool sheets and marked-up isometrics indicating the assigned spool mark numbers and any as-builting.
   - The spool sheets must be signed-off copies indicating completion of each stage of fabrication and testing.

In order to check that the fabricators understand your instruction, you should request that they send you copies of the first few spool sheets that they produce for your review and comment. Figures 5–6 and 5–7 are examples of a spool sheet.

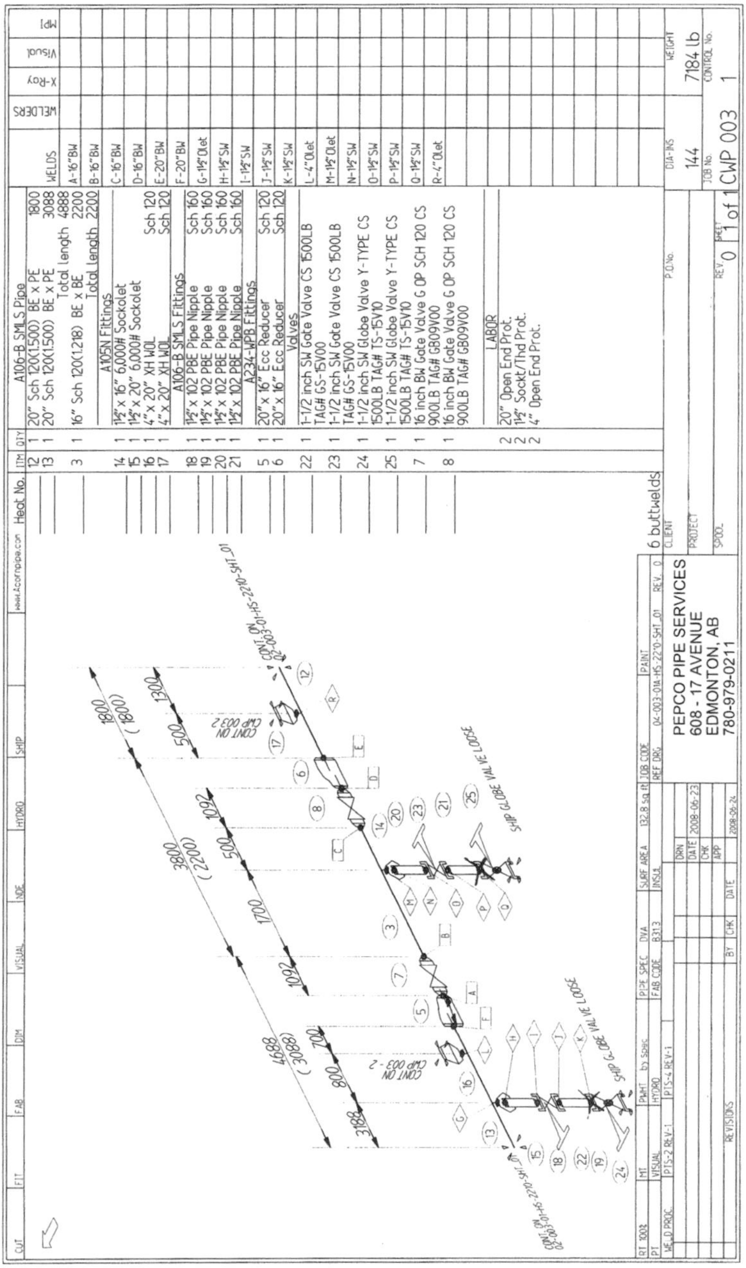

Figure 5–6 *Spool Sheet Example.* (Printed with the permission of Pepco Pipe Services, Edmonton, Canada.)

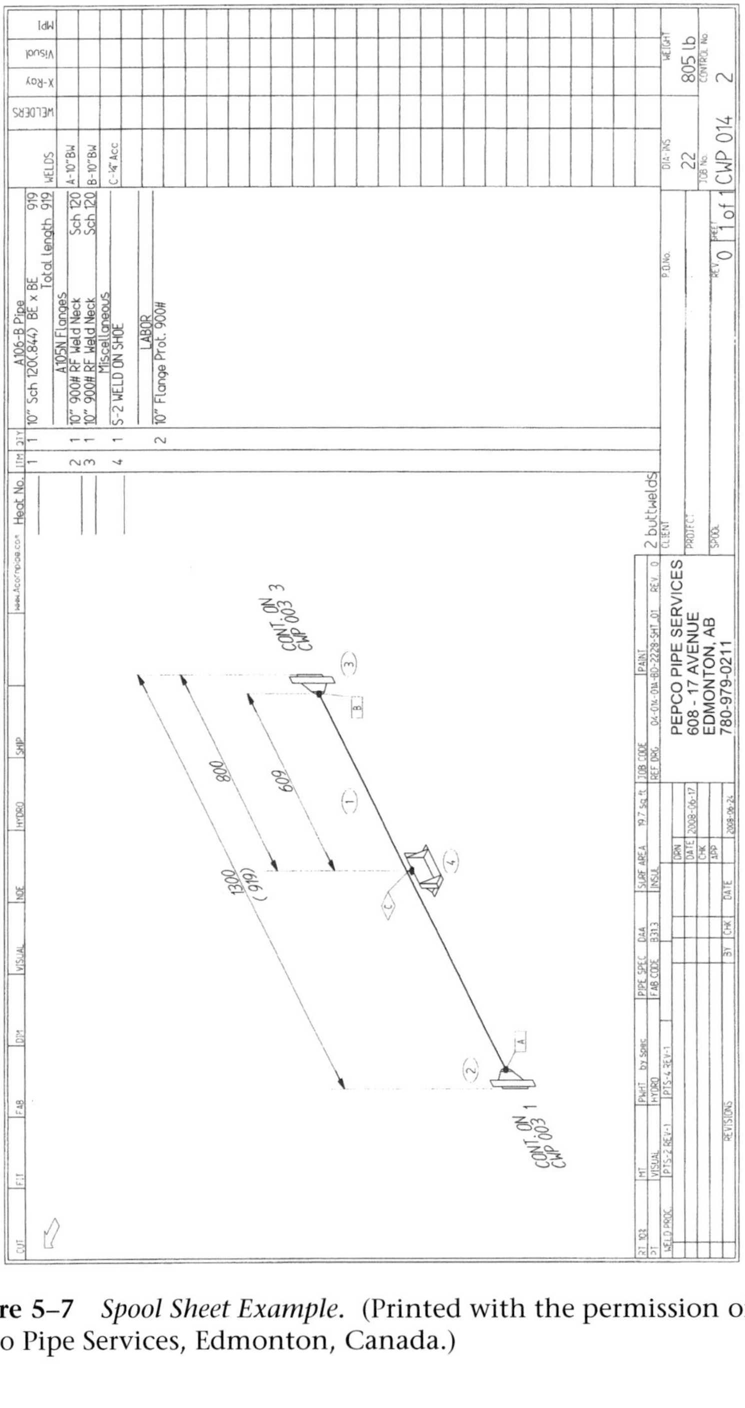

Figure 5–7 *Spool Sheet Example.* (Printed with the permission of Pepco Pipe Services, Edmonton, Canada.)

5.5 Requests For Information (RFI)

Requests For Information must be handled as a high priority. Any delays in responding to the fabricator can be a cause for delay to the fabrication schedule. Occasionally a fabricator may run into problems maintaining the schedule, and you do not want to become the target of a ready-made excuse for a delay.

5.6 Visits with the Fabricators

5.6.1 The Value of Shop Visits

Shop visits to the fabricators are very beneficial to the project. There is nothing like face-to-face meetings to build rapport and trust. You will find that the fabricator will have a whole string of questions for you each time you visit. While RFIs are a very needed and important procedure, every question does not need to go through the RFI formality. Sometimes all that is required is a small clarification, especially when people are on the right track and are only looking for some confirmation. Without these visits and the rapport that allows for a quick phone call, the only option open to the fabricator is the RFI. This can be time consuming for everyone, both on the fabricator's side and the engineering company's side. Rather than go through the delay to get an answer, the fabricator may be inclined to make an assumption.

You may wonder what you are going to talk about at visits, but topics will come up that would not have otherwise surfaced, or at the least not surfaced for some time to come.

Here are few stories for you:

> I accompanied a project engineer on the first visit to a spooling fabricator. The fabricator's project manager and drafting manager claimed in our meeting that the Material Take-Off (MTO) report received from the engineering company was deficient. The MTO report that the drafting group had compiled indicated that there was a substantial material shortage, and the project manager asked my project engineer to approve that the fabricator be allowed to purchase these deficiencies.
>
> We brought the fabricator's designer, who had compiled the MTO, into the room with the drawings he/she had used for this task. The drawings had both module and field erected piping shown on one set of piping arrangement drawings, and the reason became immediately apparent. This fabricator's

scope was for spooling of the field erected piping only. Although there was a demarcation line indicating field on the one side and module on the other, the fabricator had missed this and had done his/her material take-off for both module and field piping. The drafting group had also started to draw the shop spool sheets for all of the piping. Obviously, this would have been resolved through an RFI, but the fact that we were there on the spot resolved this quickly and without further time being wasted. It is for reasons such as this that I am so adamant about separate CWPs, models and drawings.

On another occasion at a module fabricator's shop, I was walking around observing progress and I noticed what appeared to be some overly large fillet welds on gusset plates placed within the web of the wide flange steel columns and beams of a module. When I asked about them, I was shown the shop set of the engineering company's structural standards. A simple misinterpretation of a note on a standard had the fabricator's welders over–welding every gusset plate. Instead of an 8mm fillet weld they were making the welds 16mm, which in some cases was closer to 20mm. I estimated that had this error not been caught it would have amounted to an extra $300,000 in time alone over the course of the CWP they were working on. As this was a time and materials contract, the client would have most likely ended up paying for it.

The structural engineer ordered that some NDE be conducted and everything proved to be acceptable. If repairs had been required it would have been at the fabricator's cost, however, the potential schedule delays and the associated disruptions would have been borne by the project.

On yet another occasion, at the first visit with a spooling fabricator, I was informed that the fabricator had started the fabrication and I asked to see the first batch of approximately thirty spool sheets that they had created. Somewhere along the way the fabricator's designers had misinterpreted the isometrics and other information and were detailing every field weld as a field fit-up weld with 150mm of extra pipe. When I commented that this was an error the drafting manager was so sure of his group's interpretation that he disputed my words. There was no intent to raise an RFI for clarification on this.

> Had I not been there to point out this error at an early stage it may have gone unnoticed and unchallenged until the inspector came. There would have been a schedule delay to bring all of the spools back through the shop to cut and prep the ends. I am not so sure that an inspector would have picked up on this, in which case the spools would have been in the field before the error surfaced. This error could have ended up being a significant dispute over back charges.

It is important to recognize that issues are often brought forth and resolved through an RFI, or are eventually caught through inspection or surface in some other way. The point is that you will have a history and working knowledge of the project that is very valuable in recognizing and heading off issues before they arise, but you will never know unless you are there to see the fabrication for yourself.

5.6.2 Supporting the Fabricators

The fabricators will be well structured in the moving of spools and modules through their shop, and there is likely to be little that you can help them with in this regard. Most of the people you will deal with come from a welding and construction background. The spooling and module erection expertise of the fabricators' lies primarily in knowing the fabrication side of the business, e.g., codes, regulatory requirements, welding procedures, materials, weld mapping and NDE testing, etc., for which they will have procedures for managing and tracking.

Where you will be of benefit is with your knowledge of the history of the project and your experience as a senior piping designer in general. Just as you may lack the depth of knowledge in fabrication that they have, they may lack the depth of knowledge that you possess in the designing and drafting aspects. This can make for a very mutually beneficial relationship. You may discover and correct misinterpretations in the conclusions drawn by the fabricator's own drafting team that, had they remained unchecked, would have found their way into the shop causing rework for the fabricator and delays for the project.

The first meeting you have should follow shortly after the kick-off meeting once the fabricator has had a chance to digest the contents of the CWP. The chances are that you would not attend this first meeting alone. At the least, you will have the project engineer in attendance with you because many of the questions are going to be about schedule and material deliveries; questions that you should not be answering. For subsequent meetings that become more focused, you may be on your own.

Your role in the first and subsequent meetings is to gauge the fabricator's understanding of the CWP and assist the fabricator to resolve any difficulties they are having with interpretation of:

- The engineered drawings.
- The vendor drawings.
- The standards drawings.
- The piping classes.
- The SOW.
- MTO and other reports generated from the 3-D model databases.
- CWP and 3-D model numbering convention.

There will be other ways in which you will work with them, depending on need. Some may be mandated by the project, such as double checking the spool sheets of the more expensive piping. While it is not common, on some projects it is considered prudent that you have the spool sheets checked again by your piping group for such as high pressure and alloys piping.

Some ways of working with the fabricators will occur naturally as a consequence of being there, and you may find that you start falling into a consultant role. Offering help outside of your responsibilities to the project will be a judgement call. For instance, you may be able to offer advice on a drafting or checking procedure to the drafting manager. While this would not be a requirement, it would be a friendly gesture that builds the rapport and helps the integrity of the project.

There will be times when you will have to go above and beyond the call of duty for the sake of the project. You will encounter fabricators that are struggling for one reason or another, and in the interest of keeping the project on track you may feel the need to offer some additional help to a fabricator that is falling behind, as long as those reasons are within your sphere of influence. This might be in the way of short-term help, for instance, checking to help move things along, or longer term help such as assigning someone from your team to work in the fabricator's shop. This is extremely rare, but not unheard of.

The above is applicable to fabricators whether they are utilizing 2-D CAD or 3-D CAD. However, 3-D software comes with other challenges, and if the fabricator is new to the use of 3-D, he/she may be struggling with your models and the software. Providing short-term help to assist the fabricator in CAD support may be the answer.

5.7 Automatic Spool Generation

I believe that the wave of the future is automatic spool generation. Just as engineering companies are automatically generating isometrics, likewise the fabricators will generate the spool sheets from the same models as the isometrics are generated. This is an evolution that only makes sense. When something is copied from one format to another, mistakes will happen. Working with imported files from the engineering company's models will negate the possibility of mistakes being made by the fabricator's spooling drafters due to copying information from the isometrics. Another consideration is that the engineering companies could buy the automatic spooling software and create the spool sheets as part of the CWP deliverables.

The fabricators will still have to add information to the spool sheets such as material heat numbers, NDE testing, and weld mapping, but utilizing automatic spool generation will save time and money just as automatic isometric generation has done for the engineering companies.

5.8 Conclusion

You must be aware of the limitations of your responsibility and authority. There will be times when you will be drawn into discussions concerning topics that will go beyond your mandate, such as the schedule and cost extras or engineering questions. In cases like these you must refer the fabricator to the appropriate party and remind him/her that for some issues he/she will have to use the RFI procedure. Never make promises or agreements to help that involve project hours without getting the approval of the project manager first. My previous examples of offering assistance of checking, assigning an individual from your team to assist them, and assistance from 3-D CAD support would be such appropriate times.

CHAPTER 6

Field Construction

6.1 Introduction

There are two aspects of the support provided by the piping lead to the construction management team during construction: the support from the home office and the support in the field.

6.2 Support From the Home Office

As with the support that is required for the fabricators, support from the office for the construction management team will be required to:

- Attend the kick-off meetings with the field contractors.
- Provide piping erection Scope of Work (SOW).
- Identify computers, 3-D software set-ups, and maintenance requirements.
- Respond to Requests For Information (RFI).
- Update the drawings and the 3-D models with the as-built information.

6.2.1 Kick-off Meetings

The kick-off meetings with the field contractors are like the previously discussed meetings with shop fabricators, with the exception that your attendance will be on an as needed basis. This is due to the fact that the custody and responsibilities for the completion of the work has been transferred to the construction management team from the engineering team. Unless you are assigned to the field, your involvement in the issued CWPs is more-or-less finished. This is not to say that a rapport with the field contractors is not every bit as beneficial as that with the shop fabricators, but due to the transference of responsibility you may have to be assigned to the field to have that same level of involvement.

6.2.2 Field Erection of Piping Scope of Work

We have discussed the SOWs for pipe spool fabrication and module fabrication and assembly. This last example is for the erection of the spools. It reflects the same shop and field material and fabrication splits, and material purchase splits as in the previous SOW examples.

6.2.2.1 Example of Field Erection of Piping Scope of Work

1. Scope of Work:

   Work to be performed under this Construction Work Package (CWP) No. ____________________.

   Work includes the following items, all as set forth in the drawings, specifications, and other data:

   - Install the welded spools of piping components NPS 2 and above in accordance with the attached drawings, line designation tables, piping standards, and specifications.
   - Install the lined piping and fittings, all NPS sizes, in accordance with the attached drawings, line designation tables, piping standards, and specifications.
   - Field run all NPS 1½ and below piping in accordance with the attached drawings, line designation tables, piping standards, and specifications.
   - The scope of the work is defined by the isometrics.
   - Installation dimensions are to be taken from the piping arrangements and isometrics.
   - Field welds (FW) have no trim allowance. Weld spools to exact dimensions.
   - FFWs have 150mm of additional pipe length for trim allowance. Field check the dimensions on the piping arrangements and isometrics and cut-to-suit. Prepare the BE and weld in place.

- Install the vent, drain, and pressure connection assemblies.
- Install the shipped loose shoes, guides, and anchors.
- A stub has been provided for base ell supports. Set the piping at the required elevation and complete the installation as per the standard drawings.
- Verify shim height requirements as noted and install.
- Install the field supports (miscellaneous structural steel, 'U' bolts, etc.) as detailed on the attached drawings and piping standards.
- Install all instruments not included in the spools. Primarily, these will be the inline bolted instruments.
- Install all specialty items not included in the spools. Primarily, these will be the inline bolted specialty items.
- Install all flanged valves.
- Install the pipe hangers and spring supports.
- Install the heat tracing manifolds.
- Hydro-testing will be performed by the installation contractor.
- Mark-up a set of isometrics with any as-builting, material heat numbers, and weld mapping.
- Mark-up a set of P&IDs and Line Designation Tables with any as-builting.
- Method for material requisitions by installation contractor to be determined by contract administration in conjunction with the installation contractor.
- Turnover package to be provided in whole and as defined by QA/QC requirements.

2. Work Not Included:

   The following items of work associated with this CWP do not form part of the Work of the CWP:

   - The repair of painting for any welds performed on painted pipe will not be required by the installation contractor.
   - Insulation and tracing.

3. Supplied Materials:

   The following materials will be supplied for incorporation into the work:

   - Pre-fabricated spools NPS 2 and above.
   - Pre-fabricated lined pipe and fittings, all NPS sizes.
   - Pre-fabricated heat tracing manifolds.
   - Pre-fabricated anchors, guides, clamp-on shoes, and base ell supports.
   - Flanged valves NPS 2 and above.
   - Specialty items.
   - Instruments.
   - Pipe hangers.
   - Spring supports.
   - Studs, nuts, and gaskets NPS 2 and above.
   - Blind flanges NPS 2 and above.

   Except for those items specifically listed above, the contractor will be responsible for supplying all materials required to complete the work of this CWP.

4. Schedule:

   Work of this CWP shall be performed in accordance with the following schedule:

   - Work shall commence on or about ______________.
   - Work shall be completed on or about ____________.

5. Drawings, Specifications, and Other Data

   Work shall be performed in accordance with the drawings (signed and stamped "Issued for Construction"), specifications and other data attached hereto and listed below:

There are other mechanical SOWs that we have not discussed:

- Erection of equipment.
- Installation of bridles and instruments on equipment.
- Erection of modules.
- Painting of field erected piping.

- Tracing of field erected piping.
- Insulating of field erected piping.

Beyond issuing the CWP drawing packages, it is unlikely that you will have any involvement in these SOWs because they are field activities managed by construction, and in some cases require a specialized knowledge to write.

6.2.3 Computers, 3-D Software Set-ups, and Maintenance

It has become standard practice on 3-D projects for the models to be utilized during construction. Therefore, it will be required to have computers and the 3-D software set-ups in the field with skilled designers able to run the 3-D software.
You will be involved in the following discussions:

- How many computers will be required?
- How many licences of the 3-D software will be required?
- How many licences of the 3-D viewing software will be required?
- Will there be full time CAD support in the field or will the support be by remote access from the home office?
- The procedures that will be used to interface the models with the field. You do not want the field to have access to the live models. The access that they have will be to copies of the models, commonly backed-up nightly to their server.
- How many and which of your designers will be assigned to the field to run models and provide the other needed support?

6.2.4 Request For Information

The RFIs sent during construction may originate from the construction team or the field contractors. All will be funnelled through the construction team and, as with the RFIs from the fabricators, all must be handled as a high priority. Any delays in construction can add up very quickly to a lot of money, or, conversely, can be saved by a quick response.

Significant problems will require your assistance to investigate and possibly redesign. The resources do not commonly exist in the field for the field personnel to become distracted by the time requirement of a single major problem. In these cases they may need your support so they can continue to focus their own efforts on the day-to-day activities of construction support.

6.3 Support in the Field

As the piping lead in the field, you and your team will support construction in a variety of ways. I have immensely enjoyed the times that I have worked in the field. There is an attitude that everyone can be expected to assist as required to get the job done, and you will gain experience in areas that you are not able to be involved in when working in the office. You will have a defined role, but you can also expect to do anything and everything within your capabilities that will help to solve problems and get results.

I am not suggesting that you or anyone else should ignore safety and be put in harms way. I am saying that you cannot consider any task as menial and put yourself above it. You will get along famously with the construction team if you are willing to jump in and help wherever it is needed on any given day.

Primarily, you and your team will support construction in the following ways:

- Interpreting the engineered drawings, vendor drawings, standard drawings, and the piping classes and other piping specifications.
- Reviewing the SOWs with construction and erection contractors.
- Utilizing the 3-D models.
- Setting up punch list packages.
- Compiling and submitting RFIs.
- Investigating back charges and extras.
- Progress monitoring.
- Maintaining the master stick files.
- As-builting.

6.3.1 Utilizing the 3-D Models

The 3-D models will be utilized in a number of ways:

- Generating reports from the models.
- Generating screen shots from the models.
- Construction planning.
- Problem solving.

6.3.1.1 Generating Reports

There are a number of reports that can be run from the model databases to aid construction. A few possibilities are:

- Lengths of pipe, number and types of fittings and valves in a line or system.
- Number of field welds in a line or system.
- Bulk field material to be purchased.
- Number of shoes, anchors, and guides to be installed on a line or system.

All reports are generated to assist in the construction planning and execution.

6.3.1.2 Generating Screen Shots

Screen shots are generated to assist the erectors in understanding the layout. You will find that contractors will come to you for help on the interpretation of the drawings. Sitting with them for a few minutes and going through the model of the area in question to produce a few hard copy screen shots that they can take with them will prove very valuable.

6.3.1.3 Construction Planning

The 3-D models are an indispensable tool for pre-planning the construction activities. Models provide the capabilities to isolate individual structures and piping systems, to rotate, change colors, capture screen shots and comments, and easily visualize the piping and equipment layouts. Several activities that the 3-D walkthroughs of the models assist in are:

- Identifying hazards ahead of time to create safe work plans.
- Communication between the operations start-up and commissioning team and the construction team:
    - Aids both teams to prepare for large scale shutdowns or unit outages to avoid unnecessary down time and/or costly mistakes.
    - Gives both teams a chance to review the overall project scope and agree on the best path forward to constructing and commissioning the project.

- Planning the scaffolding. Scaffolding is a large expense for any project, and the models provide the ability to plan ahead to give clear direction to the scaffolding contractors. This avoids later costly modifications and rebuilds. The screen shots that can be supplied to the scaffolding contractors also have a lot of value.
- Planning of erection sequencing.
- Planning of hydro-test packages.
- Reviewing the piping system(s) to be punch listed.

It is hard to imagine planning and executing activities with the drawings alone and without the capability of conducting a walk-through of the models.

A contractor may wish to have copies of the models to plan the work in his/her own time. There is no reason to not comply with this request. However, you should keep a record of which contractors you have given copies of the models to so that in the event of a model update from the home office, such as an FCN, you can give an updated model to the contractor as well.

6.3.2 Problem Solving

All projects will have problems during construction ranging from the significant to the minor, not because the models are inaccurate, but because errors happen in the real world. For instance, spools may have dimensional errors or piling may not be exactly placed, leading to difficulty with fit-up and other problems.

Usually your notification that there is a problem comes in the form of a contractor walking into the construction trailer with drawings in hand. Your job is to listen to the explanation of what appears to be the problem, go with the individual to investigate for yourself, take measurements to identify where the problem lies, and solve the problem. This will take the form of updating the model with the information gathered, e.g., a pump that is slightly out from the intended location, and creating an isometric sketch with the revised dimensions.

Issues that cannot be easily rectified must be addressed through an RFI.

6.3.3 Punch Lists and Deficiency Reports

When a contractor claims to have completed a piping system it is necessary to walk the lines to inspect the work and create the deficiency reports. This exercise is called punch listing, or moaning the lines. The piping designers in the field are often called upon to assemble a package of drawings for the purposes of the inspection, and sometimes to conduct the inspection.

1. Prerequisites:

   The following checks must be carried out to ensure that the piping systems have been correctly fabricated and installed. This procedure will ensure that all required rework will be carried out prior to commissioning and start-up.

   The inspector must use the latest revision of the following documents to carry out the inspection:

   - Piping classes (for reference).
   - Piping and Instrumentation Diagrams (for reference).
   - Piping standards (for reference).
   - Tracing standards (for reference).
   - Insulation standards (for reference).
   - Instrument standards (for reference).
   - Instrument vendor data sheets (for reference).
   - Specialty item vendor data sheets (for reference).
   - Equipment vendor drawings (for reference).
   - Piping arrangements.
   - Piping isometrics.
   - Tracing arrangements.
   - Tracing isometrics.

   There are three stages of inspection:

   - Pre hydro-test: intended for release of pipe system for hydro-testing.
   - Post hydro-test prior to insulating.
   - Post Insulating.

2. Inspection Checks Prior to Hydro-Testing:

   - All supports, shoes, anchors, guides, slide plates, and spring supports have been installed in the correct locations and orientations. Spring supports remain locked for the hydro-test.

- The material and rating stamped on all flanges matches the piping class.
- All bolted connections have bolts of appropriate length, i.e., two threads showing, and gaskets installed.
- All manually operated valve tags are correct and the valves are installed in the correct direction of flow where applicable.
- All manually operated valve bonnet orientations are correct.
- All check valves are installed in the correct direction of flow. Internal flappers are to be removed for hydro-testing.
- All control valves and instruments as required are not installed. In-line instruments are generally not installed at this time, as they are not tested through.
- All specialty item tags, locations, and installations are correct.
- All spectacle blinds and spades are installed in the correct locations and directions for hydro-testing, i.e., closed or open.
- All permanent or temporary strainers are installed in the correct locations and directions.
- All high point vents and low point drains are installed and comply with the standard drawings.
- All temporary shipping steel bracing or banding on modules (painted red) has been removed.

3. Inspection Checks Prior to Insulating:

- Hydro-testing is complete and witnessed for all lines.
- All manually operated valve hand wheels, stem extensions, and chain operators are installed. Chain ops and hand wheels will not necessarily be installed prior to hydro testing.
- Check valves flappers are to be re-installed after the hydro test.
- All control valve tags are correct and the valves are installed in the correct direction of flow.

- All control valve actuator orientations are correct and instrument air tubing hook-ups are completed where applicable.
- All other in-line instrument tags are correct and the instruments are installed in the correct direction of flow. Tubing to transmitters has been completed where applicable.
- Tagging and location of all on-line instruments are correct and tubing to transmitters has been completed where applicable.
- All spring supports have been unlocked.
- All spectacle blinds and spades are installed in the correct locations and directions, i.e., closed or open, or have been removed after the hydro-test.
- All painted pipes have been repaired at the field weld locations.
- All required tracing has been applied as per the tracing drawings and standards. Check that correct tube bending has been carried out with no flattening of tubing.
- For any lines that require electrical tracing, check that the tracer wiring has been applied and secured correctly.

4. Inspection Checks After Insulating:

- All lines and equipment requiring insulation, including partial personal protection, have been insulated to the correct thickness, and that cladding and banding is correct and sufficient.

The P&IDs and isometrics will be "yellowed off" during the inspections, and any discrepancies shall be clearly noted in red. The inspector is to prepare an itemized deficiency report at each stage indicating the corrective action to be taken. Submit these reports together with the marked up P&IDs and isometrics to the construction supervisor and QA/QC, and keep a copy for record.

The originals will be transmitted to the erector for rectification. Once notified that the deficiencies have been completed, the inspector will revisit the piping system and compile the final inspection report for the stage of inspection.

6.3.4 Compiling and Submitting RFIs

Clarifications may be required on a number of issues from the home office, and it may be decided to raise an RFI to log a permanent record of the discussions. But there are several occasions when an RFI must be used.

It is important to realize that neither you nor anyone in your team is in the field to redesign the piping. It is not within your mandate to alter an approved engineered design or tie yourselves up on any one exercise for an extended period of time. There are several circumstances where you may be requested or feel the temptation to do so:

- The client's operations personnel are requesting changes.
- There is a major mistake that requires a redesign. For exmaple, a piece of equipment is significantly out from the intended location.
- The design appears overly complicated and a simpler, more economical approach seems possible.
- The design does not work. For exmaple, access has been blocked.

You are in the field to assist the construction team to keep the construction moving, and you must stay focused on the immediate day-to-day issues of problem solving and recognize when you require assistance. There is a difference between problem solving that stays within the intent of the original design and making changes to an engineered design that will involve re-engineering, and may add to the scope and budget even though these changes may clearly be necessary. Small adjustments do not require a submission for approval from the project management team or the stamping engineer, whereas significant changes do. Once you start getting into areas where the changes to the IFC design involves re-engineering and redesign, manhours, scope, budget and possibly back charges and extras, an RFI is in order. This requires a judgement call at times. When you encounter situations that you are unsure of, you are advised to consult with the construction manager and the field engineer.

Gather enough information to explain the situation and offer a possible solution with a sketch, if required, and then submit an RFI to engage the assistance of the project management team. Only the stamping engineer has authority to approve changes to his/her designs and only the PMT have the authority to act on and initiate changes of this nature; to engage the necessary engineering and design team members in the office to make the design changes and re-issue the drawings.

All RFIs must be logged along with the attached sketches and a record kept of both submissions and responses.

6.3.5 Investigating Back Charges and Extras

A back charge is a charge that may be applied to a contractor to fix sub-standard work, incomplete work, or work that has errors. An extra is a charge from a contractor for an activity they claim was required to complete the work that is an addition to the scope. It is not likely that you will be involved in the contractual issues; however, you may have to investigate possible back charges and the claims for extras and compile reports to be used in discussions by those that are involved.

There are a number of reasons why adjustments may be required during the erection of a piping system so that everything will fit together:

- Spool fabrication errors.
- A module fabrication error, e.g., a connection is in the wrong place.
- Structural steel fabrication errors.
- Errors on the piling drawing or in surveying resulting in incorrect pile locations. Additional surveys may be conducted prior to equipment and module placement so that adjustments can be made via the pile caps, but this does not always happen.
- Errors in equipment manufacture, e.g., vessel saddles or nozzles may be out.
- Equipment is placed in an incorrect orientation. I worked on a project where a distillation column was placed 180 degrees out from the intended orientation.

  Equipment vendor plan drawings should always be marked-up to include the Plant North. Additionally, tanks and vertical vessels should always have the 0-degree orientation at the top of the north axis with the angles for the tank/vessel quadrants and nozzles oriented in a clockwise direction.
- Incorrect information on the piping drawings. It is sometimes the case that firm vendor data, e.g., certified face-to-face dimensions for instruments, has not been received when a CWP is due to be issued. A calculated risk may have been taken to proceed with fabrication. If "HOLDS" were not

placed on the drawings then the piping fabrication may be incorrect as a result.

It may take some time to identify the exact cause, but it cannot be assumed that because the piping does not fit that the piping is at fault.

6.3.6 Progress Monitoring

You will not be responsible for the progress, but your assistance in maintaining accurate records of the progress will be necessary. While the construction management team will establish the progress parameters to be monitored, you may have to develop the procedure for this or manage the procedure that has been established. This may take the form of setting up stick files and models of CWPs. Adding highlighting and notation on the stick files, moving drawings from an ongoing to a complete status file, and copying models with the use of color coding to identify each stage of the established progress parameters may be utilized for this purpose.

Sitting in the construction trailer and waiting for the information to come to you will not be sufficient. Progress monitoring includes going out into the field and conducting daily updates from observation.

You will be responsible for organizing your team for these activities.

6.3.7 Maintaining the Master Stick Files

Just as in the office where master stick files are maintained, master stick files are a requirement in the field also. All of the RFIs sent and received back, the FCNs received, the resolutions to any field problems and the as-builts received from the contractors should be noted on the field master stick files.

6.3.8 As-Builting

Maintaining a record of the as-builting will be an ongoing process as long as the construction is ongoing. As has been mentioned above, you can capture RFIs and FCNs on the stick files, but there will also be mark-ups from the contractors to be captured. Quite possibly there will be a number of contractors providing as-built mark-ups, and all of these must be captured in one location. The master stick files are the appropriate place to transpose the mark-ups from all the individual contractors. You may also receive the same drawing several times from the same contractor. Often the as-built P&IDs for a piping system will be turned over when the system is completed. Because the

P&IDs involve more than one system, the same P&ID may be turned over several times.

As-built prints should be stamped "AS-BUILT" and signed and dated. As-built also applies to drawings where no changes are noted, i.e., there were no changes during construction.

As-builting of the master or live models must be handled outside of the construction support activities. Because construction support is the primary reason that you and your designers are in the field, maintaining models and databases may distract from this task. Field personnel require almost immediate answers. The designers must make changes per the field information and create sketches for resolution of the problems. It is best if they keep a record of the changes, but do not work in the live models where they have to be concerned about the integrity of the databases. If a model does become corrupted they can soon request a back up from the live models.

Develop a procedure whereby copies of the master stick files are turned over to the start-up team and back to the head office for the drawings to be updated. Likewise, models must be turned over to the home office where they have the mandate and the time to copy the as-built information into the live models and ensure the integrity of the databases.

6.4 Lessons Learned

All projects should end with a wrap-up of lessons learned for future projects. During the course of the project you will become aware of the things that have gone well and the things that have not gone well or were a hindrance to the project. If for no other reason than for your own future reference, towards the end of your project take the time to make notes.

Your project management team will likely want to conduct a day of lessons learned by all the disciplines, and your input will set the tone for the many projects to follow.

6.5 Safety

We will finish this book with one last personal story and a reminder of safety.

About twenty-five years ago I was working on the construction of a gas plant. Toward the end of my tenure I decided to take some photographs of the plant. I went to a good vantage point, a distillation column approximately 45m high, and climbed the ladder to the top 360-degree platform. I stepped off the ladder onto the platform and was admiring the view while I got my camera out of my pocket. Placing the viewfinder up to my eye, I started walking while looking only through the viewfinder. I walked, took a couple of snapshots and walked again, all the time looking through the viewfinder. Then I stopped, lowered the camera and looked ahead at where I thought I wanted to go next. Right in front of me was one strand of "do not cross this line" yellow tape running from the top nozzle to the outside handrail. Beyond the tape was the continuation of the platform steel frame but no platform grating. I was standing about a foot away from a very big drop. One more step and I would have fallen through. I could have become a statistic at the age of thirty leaving two young children fatherless. I went to the site safety engineer who immediately had the ladder roped off and "no access" signs erected.

That was a personal experience on how my life could change very quickly due to not paying attention to my surroundings. I trusted that if I could go there then it was safe to be there.

Industrial facilities of any kind are full of potential hazards and I know we all agree that safety is a serious matter. Follow the rules at all times to stay safe, but remember that you have a responsibility for your own safety. Always be aware of your surroundings and your egress. You never know what may happen.

APPENDIX A

Abbreviations

Above Ground	**A/G**
Anchor Bolt	**AB**
Anchors	**A**
Bevel Both Ends	**BBE**
Bevel End	**BE**
Bevel Large End	**BLE**
Bevel One End	**BOE**
Bevel Small End	**BSE**
Bill of Material	**BOM**
Blind Flange	**BF**
Bolt Circle	**BC**
Bolt Circle Diameter	**BCD**
Bottom	**BOT**
Bottom Of Pipe	**BOP**
Butt Weld	**BW**
CAD Job Note	**CJN**
Carbon Steel	**CS**
Center Line	**CL**
Chain Operated	**CH OP**
Complete With	**C/W**
Computer Aided Design	**CAD**
Concentric	**CONC**
Construction Work Package	**CWP**
Coordinate	**COORD**

Datum	**DAT**
Design Basis Memorandum	**DBM**
Design Change Notice	**DCN**
Detail	**DET**
Diameter	**DIA**
Dimension	**DIM**
Discharge	**DISCH**
Double Extra Heavy	**XXH**
Double Extra Strong	**XXS**
Drawing	**DWG**
Eccentric	**ECC**
Elbolet	**EOL**
Electric Trace	**ET**
Elevation	**EL**
Elevation (View)	**ELEV**
Existing	**EXIST**
Expansion Joint	**EXP JT**
Extra Heavy	**XH**
Extra Strong	**XS**
Face of Flange	**FOF**
Face to Face	**F/F**
Far Side	**FS**
Field	**FLD**
Field Change Notice	**FCN**
Field Check	**FC**
Field Fit-Up Weld	**FFW**
Field Support	**FS**
Field Weld	**FW**
Figure	**FIG**
Finished Floor	**FIN FL**
Finished Grade	**FIN GR**
Fitting	**FTG**
Fitting to Fitting	**FTG/FTG**
Flange	**FLG**
Flanged	**FLGD**
Flat Face	**FF**
Flat On Bottom	**FOB**

Flat On Top	**FOT**
Flexible	**FLEX**
Floor	**FL**
Floor Drain	**FD**
Flow Line	**FL**
Foundation	**FDN**
Future	**FUT**
Gage or Gauge	**GA**
Galvanized	**GALV**
Gasket	**GSKT**
Grade	**GR**
Grating	**GRTG**
Ground	**GND**
Guide	**G**
Glycol	**GL**
Glycol Trace	**GT**
Hanger Rod	**HR**
Header	**HDR**
Heat Trace	**HT**
Heater	**HTR**
Heating, Ventilating and Air Conditioning	**HVAC**
Heavy	**HVY**
Height	**HGT**
High Point	**HP**
High Pressure	**HP**
Horizontal	**HORIZ** or **HOR**
Inside Diameter	**ID**
Instrument	**INSTR**
Instrumentation	**INSTRUM**
Insulation	**INSUL**
Inter-discipline Drawing Review	**IDR**
Iron Pipe Size	**IPS**
Issued For Approval	**IFA**
Issued For Construction	**IFC**
Issued For Engineering	**IFE**

Issued For Review	**IFR**
Kilopascal (Gauge)	**kPa(g)**
Left Hand	**LH**
Line Designation Table	**LDT**
Long Radius	**LR**
Long Radius Elbow	**LR ELL**
Low Point	**LP**
Manifold	**MAN**
Manway	**MW**
Material Take Off	**MTO**
Maximum	**MAX**
Minimum	**MIN**
National Pipe Thread	**NPT**
Near Side	**NS**
Net Positive Suction Head	**NPSH**
Nipple	**NIP**
Nominal Pipe Size	**NPS**
Non-Destructive Examination	**NDE**
Normally Closed	**NC**
Normally Open	**NO**
Not Applicable	**NA**
Not to Scale	**NTS**
Outside Diameter	**OD**
Outside Screw and Yoke	**OS&Y**
Personnel Protection	**PP**
Pipeline	**P/L**
Pipe Support	**PS**
Pipeway	**PW**
Piping & Instrumentation Diagram	**P&ID**
Piping Job Note	**PJN**
Plain Both Ends	**PBE**
Plain End	**PE**
Plain Large End	**PLE**
Plain One End	**POE**
Plain Small End	**PSE**
Plate	**PL**

Platform	**PLTF**
Post Weld Heat Treatment	**PWHT**
Pound Per Square Inch (Gauge)	**psi(g)**
Pressure Safety Valve	**PSV**
Process Flow Diagram	**PFD**
Project Execution Plan	**PEP**
Quality Assurance and Quality Control	**QA/QC**
Radius	**R**
Raised Face	**RF**
Reducer	**RED**
Reference	**REF**
Request For Information	**RFI**
Right Hand	**RH**
Right of Way	**ROW**
Ring Joint	**RJ**
Sample Connection	**SC**
Schedule	**SCH**
Scope of Work	**SOW**
Screwed	**SCRD**
Seamless	**SMLS**
Section	**SECT**
Sheet	**SHT**
Shoe	**S**
Short Radius	**SR**
Short Radius Elbow	**SR ELL**
Sketch	**SK**
Slip on	**SO**
Socket Weld	**SW**
Sockolet	**SOL**
Specification	**SPEC**
Stainless Steel	**SS**
Standard	**STD**
Station	**STA**
Steam	**STM**
Steam Trace	**ST**

Stress Relieve	**SR**
Structural	**STRL**
Suction	**SUCT**
Swage Nipple	**SWG**
Tangent	**TAN**
Tangent Line	**TL**
Thread or Threaded	**THD**
Threaded Both Ends	**TBE**
Threaded End	**TE**
Threaded Large End	**TLE**
Threaded One End	**TOE**
Threaded Small End	**TSE**
Threadolet	**TOL**
Top of Concrete	**TOC**
Top of Grating	**TOG**
Top of Steel	**TOS**
Total Installed Cost	**TIC**
Typical	**TYP**
Underground	**U/G**
Underside	**U/S**
Unless Noted Otherwise	**UNO**
Valve	**VA**
Vertical	**VERT**
Water, Oil, Gas	**WOG**
Weight or Wall Thickness	**WT**
Weld Neck	**WN**
Weldolet	**WOL**
Work Breakdown Structure	**WBS**

Index

Made in the USA
Monee, IL
12 September 2026

62977020R00152